Deep Learning for Remote Sensing Images with Open Source Software

Signal and Image Processing of Earth Observations Series

Series Editor

C.H. Chen

Published Titles

Deep Learning for Remote Sensing Images with Open Source Software
Rémi Cresson

Sea Ice Image Processing with MATLAB®
Qin Zhang and Roger Skjetne

Compressive Sensing of Earth Observations
C.H. Chen

Radar Imaging for Maritime Observation
Fabrizio Berizzi, Marco Martorella, and Elisa Giusti

Principles of Synthetic Aperture Radar Imaging
A System Simulation Approach
Kun-Shan Chen

Remote Sensing Image Fusion
Luciano Alparone, Bruno Aiazzi, Stefano Baronti, and Andrea Garzelli

Deep Learning for Remote Sensing Images with Open Source Software

Rémi Cresson

CRC Press
Taylor & Francis Group
Boca Raton London New York

CRC Press is an imprint of the
Taylor & Francis Group, an **informa** business

CRC Press
Taylor & Francis Group
6000 Broken Sound Parkway NW, Suite 300
Boca Raton, FL 33487-2742

Printed on acid-free paper

ISBN: 978-0-367-85848-3 (hbk)

ISBN: 978-0-367-51898-1 (pbk)

ISBN: 978-1-003-02085-1 (ebk)

Visit the Taylor & Francis Web site at
http://www.taylorandfrancis.com

and the CRC Press Web site at
http://www.crcpress.com

Contents

Preface

The volume of earth observation data has increased in recent years, and more and more remote sensing imagery is available today. On one hand, the amount of remote sensing data is constantly growing due to the rise of very-high-resolution sensors and short repeat cycle satellites. Thanks to programs like Landsat or Sentinel, many free imagery products are available. In addition, community-based geographic information gathering systems are expanding each in existing geospatial data bases: the OpenStreetMap initiative is a well-known example. Regarding this amount of geospatial data, tackling complexity in earth observation information extraction is rising as a major and exciting challenge.

Deep learning is a growing trend in big data analysis, and had a breakthrough impact in the last few years on such diverse domains as image analysis, speech recognition, autonomous cars, or the arts. Convolutional Neural Networks are designed to extract features in images, enabling image recognition, object detection, and semantic segmentation. Recurrent Neural Networks are suited for sequential data analysis, such as speech recognition and action recognition tasks. As for numerous fields, the deep learning revolution has impacted the remote sensing sector. In recent years, a number of studies have shown that remote sensing benefits strongly from these new approaches, thanks to the availability of data and computing resources. Deep learning allows researchers and engineers in remote sensing to move beyond usual approaches and tackle a number of problems with solid results. Various remote sensing problems have been successfully addressed with deep learning: classification, segmentation, object detection, image restoration, image enhancement, etc. Deep learning has proven to be pertinent in many kinds of remote sensing imagery: synthetic aperture radar, hyperspectral imagery, very-high-resolution images, time series, etc.

A lot of the literature explains deep learning theory, but this book rather focuses on its technical application on satellite imagery. The core of this work teaches how to apply deep learning techniques to real-world remote sensing images using existing, open-source tools like QGIS, TensorFlow and Orfeo ToolBox. The generation of land cover maps from large satellite images will be the central topic. However, perspectives of applying deep learning to various earth observation sensors for a number of purposes are open, and we will show that deep learning applies not only to image classification, but also to image restoration from multimodal imagery.

After a short summary of deep learning background, the book will introduce the common steps to extract samples from remote sensing images, create and train deep networks, and use them to generate output images, e.g. land cover maps. Various approaches and deep network architectures will be introduced in different parts of the book. For each of them, all steps enabling the reader to perform the processing of data will be detailed. We provide an online repository containing the ancillary data and sharing code snippets used in the exercises. The software involved in this tutorial are open-source, and instructions to install them are provided.

Author

Rémi Cresson received an electrical engineering degree from the Ecole Nationale Superieure de l'Energie, de l'Eau et de l'Environnement, Grenoble Institute of Technology, France, in 2009. He is now with the French National Institute for Agricultural Research, in the Land, Environment, Remote Sensing and Spatial Information Joint Research Unit at the University of Montpellier, France. His research and engineering interests include geospatial image processing at scale, high-performance computing, machine learning, and geospatial data interoperability. He is involved in open source software development, a member of the Orfeo ToolBox Project Steering Committee, and also a member of the open source geospatial foundation OSGeo.

Part I

Backgrounds

1

Deep learning background

In this section, we provide the essential principles of deep learning. After reading this chapter, one should have the required theoretical basis to understand what is involved in processing geospatial data in the rest of the book.

1.1 What is deep learning?

Deep learning is becoming increasingly important to solve a number of image processing tasks [1]. Among common algorithms, Convolutional Neural Network- and Recurrent Neural Network-based systems achieve state-of-the-art results on satellite and aerial imagery in many applications. For instance, synthetic aperture radar (SAR) interpretation with target recognition [2], classification of SAR time series [3], parameter inversion [4], hyperspectral image classification [5], anomaly detection [6], very-high-resolution image interpretation with scene classification [7, 8], object detection [9], image retrieval [10], and classification from time series [11]. Deep learning has addressed other issues in remote sensing, like data fusion (see [12] for a review) e.g. multimodal classification [13], pansharpening [14], and 3D reconstruction.

Deep learning refers to artificial neural networks with *deep* neuronal layers (i.e. *a lot of* layers). Artificial neurons and edges have parameters that adjust as learning proceeds. Inspired from biology, an artificial neuron is a mathematical function modeling a neuron. Neuron *parameters* (sometimes called *weights*) are values that are optimized during the training step. Equation 1.1 provides an example of a basic artificial neuron model. In this equation, X is the input, y the output, A the values for the *gains*, b is the *offset* value, and f is the activation function. In this minimal example, the parameters of the artificial neuron are *gains* and one *offset* value. Gains compute the scalar product with the input of the neuron, and the offset is added to the scalar product. The resulting value is passed into a non-linear function, frequently called the *activation function.*

$$y = f(AX + b) = f[\sum_i (a_i \times x_i) + b] \tag{1.1}$$

Weights modify the strength of the signal at a connection. Artificial neurons may output in non-linear functions to break the linearity, for instance, to make the signal sent only if the resulting signal crosses a given threshold. Typically, artificial neurons are built in layers, as shown in figure 1.1.

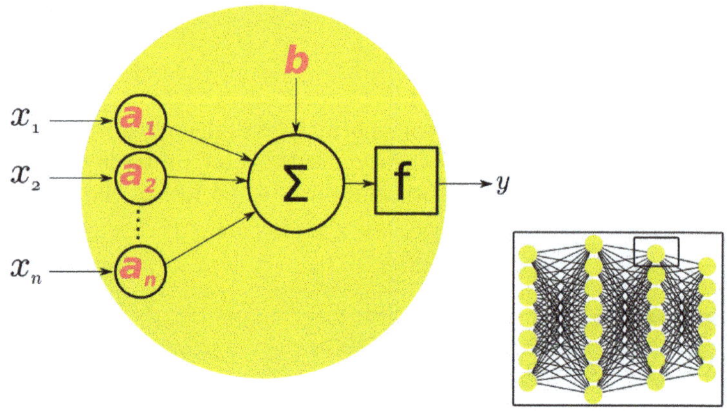

FIGURE 1.1
Example of a network of artificial neurons aggregated into layers. In the artificial neuron: x_i is the input, y is the neuron output, a_i are the values for the gains, b is the offset value, and f is the activation function.

Different layers may perform different kinds of transformations on their inputs. Signals are traversing the network from the first layer (the input layer) to the last layer (the output layer), possibly after having traveled through some layers multiple times (this is the case, for instance, with layers having feedback connections). Among common networks, Convolutional Neural Networks (CNNs) achieve state-of-the-art results on images. CNNs are designed to extract features in images, enabling image recognition, object detection, and semantic segmentation. Recurrent Neural Networks (RNNs) are suited for sequential data analysis, such as speech recognition and action recognition tasks. A review of deep learning techniques applied to remote sensing can be found in [15].

In the following we will briefly present how CNN works. Basically, a CNN processes one or multiple input multidimensional arrays with a number of operations. Among these operations, we can enumerate convolution, pooling, non-linear activation functions, and transposed convolution. The input space a CNN "sees" is generally called the **receptive field**. Depending on the implemented operations in the net, a CNN generates an output multidimensional array with output size, dimension, and physical spacing that can be different from the input.

1.2 Convolution

Equation 1.2 shows how the result Y of a convolution with kernel K on input X is computed in dimension two (x and y are pixel coordinates). In this equation, Y and K are 2-dimensional matrices of real-valued numbers.

$$Y(x,y) = (X * K)(x,y) = \sum_i \sum_j X(x-i, y-j) \times K(i,j) \tag{1.2}$$

Because of the nature of the convolution, one CNN layer is composed of neurons with edges that are massively shared with the input: it can be assimilated to a small densely connected neural layer that performs the whole input locally (i.e. with connections to input restricted to a neighborhood of neurons). Generally, CNN layers are combined with non-linear activation functions and pooling operators. The convolution can be performed every Nth step in x and y dimensions: this is usually called the *stride*. The output of a convolution involving strides is described in equation 1.3, where s_x and s_y are the stride values in each dimension.

$$Y(x,y) = (X * K)(x,y) = \sum_i \sum_j X(x \times s_x - i, y \times s_y - j) \times K(i,j) \tag{1.3}$$

A convolution can be performed keeping the valid portion of the output, meaning that only the values of Y that are computed from every value of K over X are kept. In this case, the resulting output has a lower size than the input. It can also be performed over a padded input: zeros are added all around the input so that the output is computed at each location, even where the kernel doesn't lay entirely inside the input. This is generally used to keep the input size through the convolution. However, the borders of the output are affected by the zeros of the padding. Figure 1.2 shows how the stride and padding of the convolution affect the output size (the number of rows and columns of the resulting image) and the physical spacing (the physical size of the output pixel).

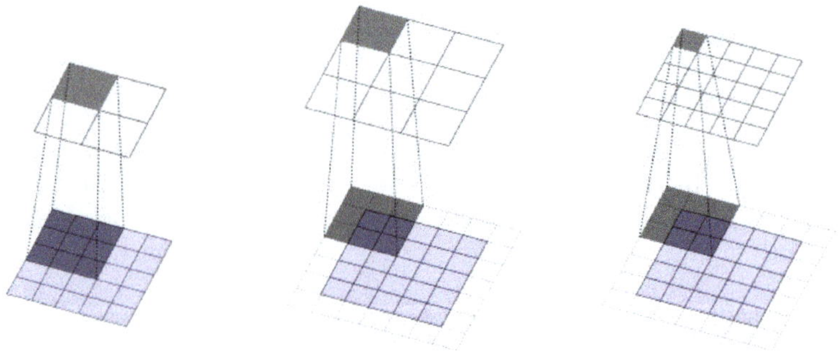

FIGURE 1.2
Left: no padding, stride=2, **Center**: padding, stride=2, **Right**: padding, stride=1. The output is the upside, white matrix, and the input is the blue matrix at the bottom. The output **size** depends on the following parameters of the convolution: padding, kernel size, and stride. The output **physical spacing** is changed accordingly to the stride in the convolution: n strides ensure that physical spacing is scaled by n.

Important note

- Non-unitary strides modify the output physical spacing.
- Without padding, the output size is changed, depending on the kernel size.
- With padding, the output size can be kept (same as the input, if no stride) but values at the border might be contaminated by the zero padding.

1.3 Pooling

Another important concept of CNNs is pooling, which is a form of non-linear downsampling. Several non-linear functions can be implemented in pooling operations, among which max pooling is the most common: it partitions the input image and for each sub-region, outputs the maximum ref to figure 1.3. The intuition is that the exact location of a feature is less important than its rough location relative to other features. The pooling layer aims to progressively reduce the spatial size of the representation, to reduce the number of parameters and amount of computation in the network, and hence to also control overfitting. It is common to periodically insert a pooling layer between successive convolutional layers in a CNN architecture. The pooling operation provides another form of translation invariance. The pooling layer operates independently on every depth slice of the input and resizes it spatially. Like convolution with non-unitary stride, the pooling operators also modify the output size.

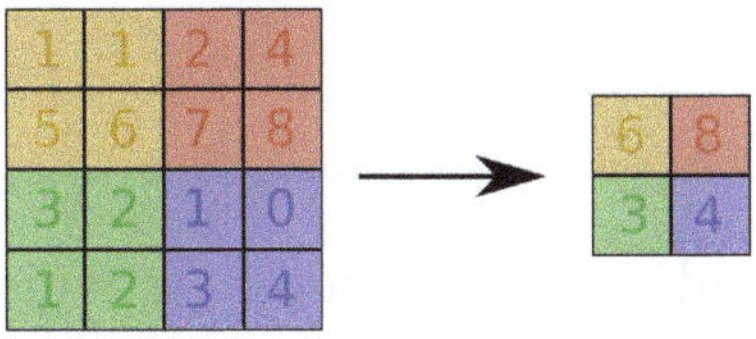

FIGURE 1.3
2×2 max pooling with stride 2. Each 2×2 sub-region of the image, the maximum value is kept.

> **Important note**
> Pooling, and convolutions with stride, can be viewed as a subsampling process, which does modify the output size, and the output physical spacing. Depending on the implementation, it can also keep partially sampled items at borders.

1.4 Activation functions

Inspired from biology, the activation functions are usually an abstraction representing the way a neuron will be activated (i.e. it will transmit its output value) or not. Basically, the activation functions control the amount of what is actually transmitted to output. There are many activation functions used in deep learning. Among these, we can enumerate the following, which are common ones we will use during this tutorial:

1. The rectified linear unit returns a maximum value between 0 and x:
$$f(x) = max(0, x)$$
2. A leaky rectified linear unit allows a small, positive gradient when the unit is not active. It is parametrized with α, a scalar value:
$$f(x) = \begin{cases} x & \text{if } x > 0, \\ ax & \text{otherwise,} \end{cases}$$
3. Sigmoid:
$$f(x) = \frac{1}{1 + e^{-x}}$$

4. Hyperbolic tangent (Tanh), which is a shift-scaled sigmoid ($tanh(x) = 2 \times sigmoid(2x) - 1$):

$$f(x) = \frac{e^x - e^{-x}}{e^x + e^{-x}}$$

1.5 Challenges ahead for deep learning with remote sensing images

There are still challenges remaining ahead to process real-world remote sensing images with deep learning. One crucial point is software implementation. Entering into deep learning requires familiarity with a framework. Implementing new methods requires an extensive programming background and deep learning knowledge. The tools must scale enough to the huge size of real-world remote sensing images and datasets. In particular, the image size problem is out of deep learning frameworks scope. In computed vision, popular problems and datasets are often related to small-sized and true-color images. Many deep learning architectures must be redesigned to suit remote sensing images. Another point is that algorithms presented in the literature often serve one unique purpose and each new algorithm implementation requires new effort and expertise to be applied on standardized geospatial data. Additionally, recent studies have pointed out that using the features computed with deep nets in machine learning algorithms like Support Vector Machines or Random Forests, offers great opportunities. However, even if most remote sensing image processing software implement a machine learning framework with such algorithms, only a few of them enable the combination of deep learning algorithms with those already implemented well-known machine learning algorithms.

In the next section, we introduce the software that leverages all these issues.

2

Software

Software is a crucial point in both remote sensing and deep learning. This section introduces the open source software involved in the practice sessions of the book.

2.1 Orfeo ToolBox

There are many types of geospatial images processing software, of which many are powerful open source programs like GRASS, SAGA, GDAL, and the **Orfeo ToolBox**.[1] The Orfeo ToolBox, also known as **OTB** ref to figure 2.1, is an open source library for remote sensing image processing developed by the French spatial agency (CNES). It contains about a hundred user-oriented applications including geometric and radiometric corrections, image manipulation, statistics, etc. It is suited for large-scale remote sensing image processing, offering a great number of functions and algorithms [16]. It is based on the Insight ToolKit[2] (ITK), a widely used open source library for medical image computation, and the Geospatial Data Abstraction Library[3] (GDAL). Additional optional libraries enrich the features of OTB (OpenCV, Shark ML, etc.).

FIGURE 2.1
The Orfeo ToolBox.

[1] `https://www.orfeo-toolbox.org/`
[2] `https://itk.org/`
[3] `https://gdal.org/`

2.1.1 Applications

The Orfeo ToolBox applications aim to provide users a number of implemented algorithms. These applications can be used directly through multiple interfaces like command line, graphical user interface, and third-party software like QGIS for instance. For developers, applications can also be used from Python and C++ APIs. Each OTB application comes with a set of parameters. Parameters used in OTB applications are typed (images, vector data, file, folder, integer, floating, etc.) and this formalism makes the OTB applications usable through all interfaces and APIs.

2.1.2 Streaming mechanism

The streaming mechanism enables the processing of images several times on different regions. Instead of working on the entire data, which would require a lot of memory, blocks are processed one by one until the output data is produced. This mechanism preserves the memory footprint and enables the processing of very large data. Most of the algorithms implemented in OTB applications and C++ classes support this mechanism. The few algorithms that don't implement the streaming mechanism require that all input and output data are stored in memory at execution.

2.1.3 Remote modules

For developers, OTB eases the integration and the deployment of new features thanks to the external modules. This mechanism enables you to standardize the development of components external to the library, such as a new OTB application. Remote modules allow anyone to extend the functionalities of OTB without being part of the core project repository. Remote modules can have different licenses than the main OTB repository. For instance, the OTBTF remote module that we will use in this tutorial, implements deep learning using the TensorFlow library. It is not in the core of the official OTB library, but provides applications that can be used as any other using the OTB application API.

2.1.4 The Python API

The Python API of the Orfeo ToolBox enables you to build pipelines using in-memory connection of existing OTB applications. The resulting pipeline still enforces the streaming mechanism. Another useful feature is the ability of using Numpy arrays in pipelines, as input or output of the OTB applications. The Python API is really great to build complex pipelines with already existing algorithmic blocks like the OTB applications (including the applications from remote modules). This book targets user-oriented interactions with OTB applications, and Python bindings are partially introduced in section 16.3.

We invite curious readers who want to exploit the full potential of the toolbox using the Python API, to read the official documentation along with the available tutorials online. Additional help can be obtained from the users community on the dedicated forum from the Orfeo ToolBox website.[4]

2.2 TensorFlow

TensorFlow is the high-performance numerical computation library for training and inference on deep neural networks [17], initiated by Google. It has become a popular open source deep learning framework, with a large worldwide growing community of developers ref to figure 2.2.

FIGURE 2.2
The TensorFlow library.

2.2.1 APIs

TensorFlow has a powerful high-level Python API to build deep nets. Currently, TensorFlow uses Keras which allows you to build networks from a higher implementation level. TensorFlow also has a simple yet powerful C++ API to integrate deep nets in C++ projects.

2.2.2 Computations

TensorFlow uses symbolic programming that distinguishes definitions of computations from their proper execution. In TensorFlow, tensors are abstraction objects of the operations and values in the memory, simplifying manipulation regardless of the computing environment: for instance, Central Processing Unit (CPU) or Graphical Processing Unit (GPU). In TensorFlow, the so-called *model* consists of operations arranged into a graph of nodes. Each node is an operation taking zero or more tensors as inputs, and producing one or multiple tensors. This data flow graph defines the operations (e.g. linear

[4]`forum.orfeo-toolbox.org`

algebra operators), and the actual computations are performed in the TensorFlow *session*.

2.2.3 Graphs

The Python API enables the construction of TensorFlow graphs, and the session runs the graph, delegating calculations to low level, highly optimized routines. Among tensors, we can distinguish concepts such as *Placeholders*, *Constants* and *Variables*. A Placeholder is a symbol hosting input data, e.g. a set of images. As its name indicates, Constants are tensors with constant values, and Variables hold non-persistent values, e.g. parameters to estimate during training. Variables must be explicitly initialized, and can be saved or restored during a session along with the graph.

2.3 Orfeo ToolBox + TensorFlow = OTBTF

Nowadays several popular open source frameworks exist, such as TensorFlow, Theano or PyTorch, but using them requires an extensive programming background and deep learning knowledge, which is limiting deep learning democratization in the remote sensing community. This book intends to introduce deep learning applied on geospatial data, even if the reader is not a Python coder and signal processing expert. The OTBTF remote module of the Oreo ToolBox will be used to support the interaction between the remote sensing and the deep learning worlds. Thanks to the user-friendly interface provided by OTB, using geospatial data as input for deep networks is eased. OTBTF uses TensorFlow internally to bring the deep learning magic acting with the same interface as OTB, but not in refraining deep learning experts from using their own model. The main features of OTBTF are the following:

1. It is open source and cross-platform.
2. It allows users without programming skills to use deep nets on remote sensing images.
3. As an extension of the Orfeo ToolBox, it inherits from the following advantages:
    - The powerful API of OTB (available interfaces, applications chaining).
    - Seamless interactions with the existing machine learning and geospatial data sampling framework of OTB.
    - It implements the streaming mechanism (see section 2.1.2) and thus allows the processing of very large remote sensing images

in a computationally efficient manner (benchmarks are available in [18]).

4. Deep learning experts can still use their coding skills to build/train their models using their favorite API (e.g. Python) and use the ***SavedModel*** in OTBTF to produce their resulting geospatial images.

2.3.1 Installation

Still, the OTBTF remote module is not currently integrated in the official OTB release. Currently, there are several ways to install it:

- Use the ready-to-use environment in the provided VirtualBox instance. The only required software for this tutorial is the Oracle VirtualBox software, that can be freely downloaded online.[5] It is the easiest option if you want to give a try. However, please note that this is not optimized for performance. You can contact the author to request the download of the virtual machine.
- Download the official OTBTF Docker image on DockerHub.[6] This option is great for Linux and Mac users since Docker efficiently uses computer resources. If you have an NVIDIA GPU you can choose the Docker image supporting the NVIDIA runtime.
- Compile OTBTF from the source. For this, please follow the instructions provided on the OTBTF GitHub repository. This is the difficult way, but you will be able to build the software with the appropriate optimization flags for your hardware.

If any problems are encountered during the installation, we refer the reader to the up-to-date troubleshooting section of the OTBTF repository. Further questions can also be asked on the Orfeo ToolBox forum[7] and tickets opened on the OTBTF GitHub repository.

Recommendations
We strongly advise Linux and Mac users to use Docker images for better performance. Another important point is the hardware: Graphical Processing Units (GPUs) drastically speed up the processes in deep learning. In particular, for chapter 14 we recommend using a GPU to run the model training. The OTBTF Docker image supporting nvidia-Docker provided in the OTBTF GitHub repository simplifies greatly the use of

[5] `www.virtualbox.org`
[6] `https://hub.docker.com/u/mdl4eo`
[7] `forum.orfeo-toolbox.org`

the software on GPU-enabled hardware. Instructions are available in the documentation section of the OTBTF repository.

2.3.2 Featured applications

The remote module contains a set of new user-oriented applications. Among others, the key applications that will be used in this tutorial are the following:

- **PatchesSampling** is dedicated to extraction of patches in multiple images.
- **TensorflowModelTrain** can train a TensorFlow model from multiple input images.
- **TensorflowModelServe** can run a TensorFlow model on multiple input images.

There is also a set of composite applications, that chain together existing OTB applications and the above applications to perform state-of-the-art machine learning (e.g. combining Random Forest with features from deep learning). In the practice session, we will see how to use these new applications with remote sensing images.

2.3.3 Principle

As explained in section 2.2, TensorFlow provides APIs to allow developers to easily build deep nets. Once a deep net is built, it can be exported as a ***SavedModel***,[8] serialized as a Google Protobuf (language-neutral, platform-neutral, extensible mechanism for serializing structured data). This ***Saved-Model*** includes the processing graph, and the variables of the model. It can then be used in the new OTB applications provided by the OTBTF remote module. OTBTF implements mechanisms to convert images into tensors that can feed the TensorFlow models, execute the model, and convert back the resulting tensors of the model into output images. Thanks to the streaming mechanism of OTB, there is no limitation on processed image size. A Tensor-Flow model is associated with its following intrinsic parameters, that must be given to OTBTF:

1. **The receptive field**: as explained in section 1, the receptive field is the input volume that the model "sees" in the input space.
2. **The expression field**: it describes the volume that the model "creates" in the output space.
3. **Scale factor**: the scale factor describes the physical spacing change between one reference input (typically, the first input) and the output space. For instance, we consider a CNN that transforms a single

[8]www.tensorflow.org/guide/saved_model

input into a single output, implementing a total of 2 pooling operators with stride 2. Then the total scale factor is 4, meaning that the output spacing is multiplied by a factor of 4, and hence the output image size will be divided by 4 in each x and y dimension.

The **TensorflowModelServe** application automatically computes the region propagation in the pipeline (i.e. from input images to output), and the output image information, which are two crucial steps to enable streaming. Figure 2.3 illustrates how related the above deep nets parameters are in these steps.

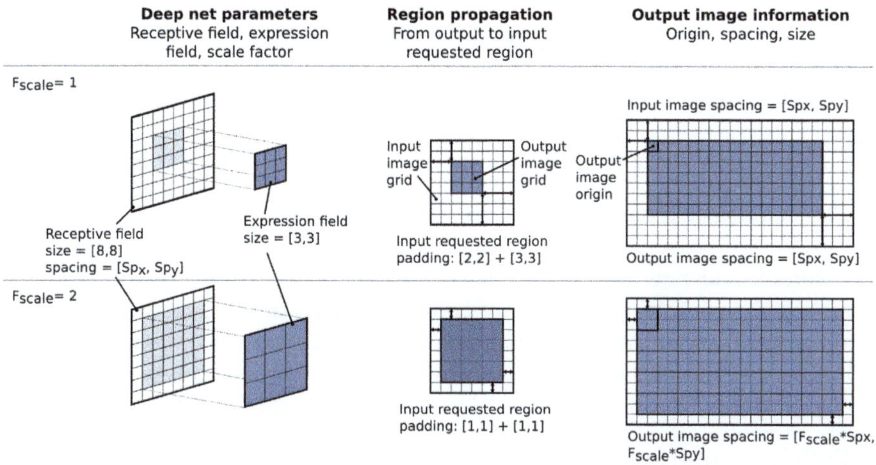

FIGURE 2.3
Deep net parameters. The figure shows the incidence of two deep net parameters on region propagation and output image information. The two nets have the same Expression Field and Receptive field, but a different Scale Factor parameter ($F_{scale} = 1$ for the net on top, $F_{scale} = 2$ for the net on bottom).

The application computes regions of images that are aligned to the expression field, i.e. image regions which have a start index and a size that are a multiple of the expression field. When a requested output region lies across multiple aligned sub-regions, the TensorFlow model is run over the largest aligned region, and only the requested part is then kept (figure 2.4). This mechanism avoids blocking artifacts and ensures that the output is independent of the tiling layout used by the streaming manager, i.e. (i) seamless, if allowed by the model, and (ii) reproducible.

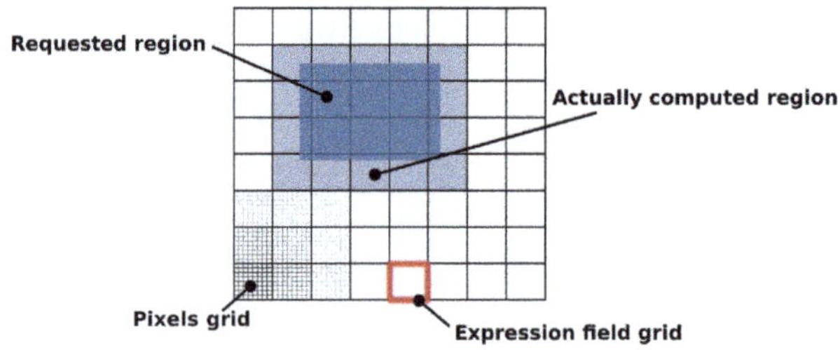

FIGURE 2.4
Computed regions alignment. To guarantee that the output is seamless and reproducible, the requested region is computed from the region which is included in the larger region aligned to the grid formed by the expression field of the net.

The **TensorflowModelServe** application can run deep nets in two modes:

- **Patch based**: Extract and process patches independently at regular intervals. Patch sizes are equal to the receptive field sizes of inputs. For each generated output image region, tensors of patches feed the TensorFlow model, then resulting tensors are translated back to the output image. This approach is straightforward, but not efficient in term of processing for large images because of the numerous transformations between images and tensors of patches.

- **Fully convolutional**: Unlike patch-based mode, it allows the processing of an entire requested region. For each generated output image region, tensors composed of one single element, corresponding to the input image requested region, is fed to the TensorFlow model. This mode requires that receptive fields, expression fields and scale factors are consistent with operators implemented in the TensorFlow model, and input images' physical spacing and alignment. This approach is computationally efficient since the transformations between images and tensors of patches are avoided: the network is applied directly on the input images.

2.3.4 Multiple input sources and outputs

The applications provided in OTBTF are designed to process any number of input image sources. This is motivated by the fact that data fusion is a challenge that deep learning can leverage easily, using deep nets tailored to one specific problem. Today's revolution in remote sensing is the high availability of optical and SAR time series, thanks to the Sentinel and Landsat programs. In addition, very-high-resolution sensors are sharper in each generation and nowadays there are a lot of sensors with sub-metric pixel spacing (Pleiades, TerraSAR, etc.). Regarding this matter, deep learning typically allows you to design a deep net that inputs time series and very high remote sensing images to generate one land cover map [13]. In OTBTF, an **input source** consists of the following input parameters:

- An **image list**: it can be either one image or a list of images (when a list of images is set, images are stacked together in their order of appearance, and hence must have the same size and spatial extent).
- The **receptive field**: size of the receptive field of the network, related to the input, in x and y direction.
- A **placeholder** name (string): as named in the TensorFlow model.

For each input source, the placeholder with the name specified by the **placeholder** parameter, is fed to the TensorFlow model with a chunk of the **image stack** that corresponds to the **receptive field** of the input.

In the same way, multiple outputs of a TensorFlow model can be retrieved as a result. The following parameters must be provided by the user to transform the output tensors into the output image:

- The output tensors names: a list of strings representing the name of the tensors, as named in the TensorFlow model (parameter **output.names**).
- The output expression field of the output tensor (parameters **output.efieldx** and **output.efieldy**).
- The output scale factor, related to the first input source (i.e. "**source1**") physical pixel spacing (parameter **output.spcscale**).

Multiple output tensors can be fetched, stacked, then exported as an OTB image. The principle is that the user provides the list of tensor names (**output.names**) to be considered as outputs of the model. After the TensorFlow session run, they are stacked together in the 3rd dimension of the output image (i.e. pixel components). Currently, only one output image can be generated. Hence, the current limitation is that the output tensors can be stacked as mentioned.

2.4 QGIS

QGIS (was also known as Quantum GIS), is a free and open source (GNU GPLv2) ref to figure 2.5 cross-platform geographic information system application that supports viewing, editing, and analysis of geospatial data.[9] Since 2002, it has grown with an increasing number of contributors and users. We will use QGIS to visualize all geospatial data used and produced during the

[9] `www.qgis.org`

tutorial. An interesting feature of QGIS is the Python extension provider. It provides access to a number of plugins written in Python, that use the QGIS API to provide new functionalities. In particular, we will use a plugin which allows us to download OpenStreetMap data as vector layers, **OSM Downloader**.

FIGURE 2.5
QGIS.

Part II

Patch-based classification

3

Introduction

This chapter introduces what we call *patch-based classification* for land cover mapping. This task consists of classifying image patches of fixed size, and attributing the estimated class to the central pixel of the patch to produce a land cover map. This kind of approach is suitable when the available data is sparsely annotated: the only requirement is the location of patches of which the class is known. Typically, examples of terrain truth allowing this approach can be GPS coordinates associated with land cover class, or manually annotated polygons in GIS, etc. This approach involves deep networks that input a patch of image, and produce a single value that represents the class of this patch: hence training this kind of network requires only input image patches, and associated class values. Once the training of a network is done, producing land cover maps consists of applying the model on each patch of a remote sensing image, and storing the value of the estimated class in the output image. The advantage of this kind of approach is its suitability to sparsely annotated datasets. Its implementation subtleties and limitations will be discussed in this chapter.

4

Data used: the Tokyo dataset

This section aims to present the data involved in this practice session. We rely on freely available data, called the *Tokyo* dataset. In the following part, we explain how to download and prepare this dataset in order to be ready for the practice session. Create a directory named *tokyo* on your file system. We will use this directory to store all the downloaded and generated data during the exercises.

4.1 Description

The *Tokyo* dataset consists of one Sentinel-2 image and one rasterized OpenStreetMap[1] (OSM) vector layer. It will be used for all exercises of Part II.

4.2 Remote sensing imagery

We use a Sentinel-2 image acquired over the Japan mainland, in the area of Tokyo (figure 4.1).

[1] `www.openstreetmap.org`

FIGURE 4.1
Sentinel-2 image acquired over Tokyo (Copernicus Open Access Hub website).

To download the product at no cost, you can use the European Space Agency hub,[2] then search for the product with the following attributes:

- Area: Tokyo (You can zoom over the area of Tokyo, as shown in figure 4.2, and draw a region of interest on the map)
- Sensing date: from *2019/05/08* to *2019/05/08*
- Mission: *Sentinel-2*
- Platform: *S2A_**

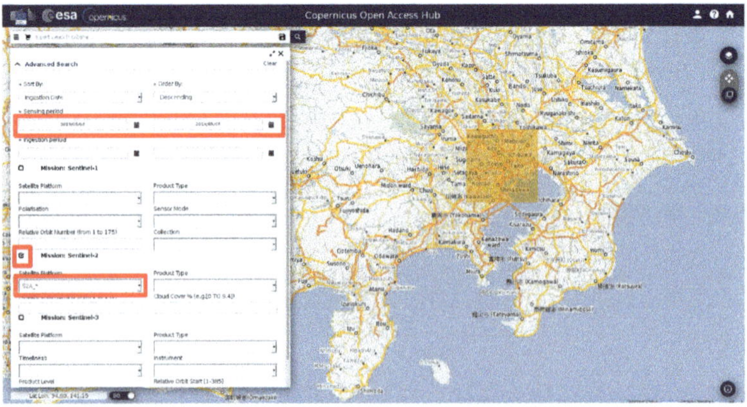

FIGURE 4.2
Enter the following attributes to search the Sentinel-2 product: Sensing date from *2019/05/08* to *2019/05/08*, Mission: *Sentinel-2*, Platform: *S2A_**. Then select the area of Tokyo and validate the search.

---

[2] `https://scihub.copernicus.eu/dhus/#/home`

From the search result, you should be able to download a product named **S2A_MSIL2A_20190508T012701_N0212_R074_T54SUE_20190508T041235** as illustrated in figure 4.3.

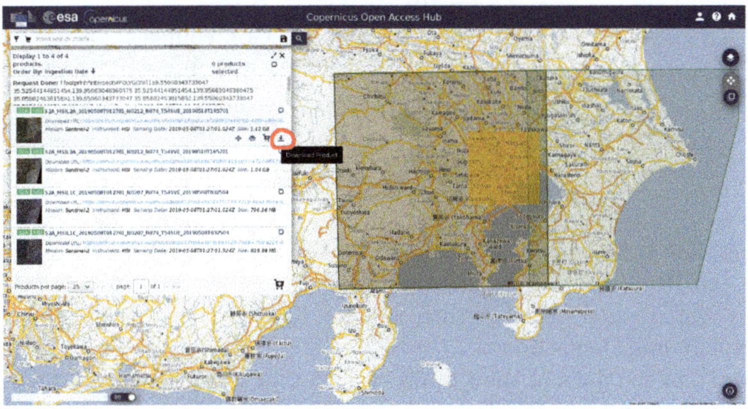

FIGURE 4.3
Download the product named **S2A_MSIL2A_20190508T012701_N0212_R074_T54SUE_20190508T041235**.

Once the product has been downloaded, you can unzip the archive and look into the extracted files. The product has two sets of images: one has 10-m spacing (spectral bands 2, 3, 4 and 8) and the other has 20-m spacing (spectral bands 5, 6, 7, 8a, 11 and 12). We won't use the other spectral bands in this chapter since they are mostly related to atmospheric contents. Table 4.1 summarizes the images' characteristics.

4.3 Terrain truth

We will use some terrain truth provided by the TETIS research facility (University of Montpellier, France). Terrain truth consists of two rasters of labels,

TABLE 4.1
Characteristics of the images

Id	Product type	Image size ($col \times row$)	Values encoding	Pixel spacing (meters)
Sentinel-2 (bands 2,3,4,8)	Multispectral	10980×10980	int (16 bits)	10 m
Sentinel-2 (5,6,7,8a, 11, 12)	Multispectral	5490×5490	int (16 bits)	20 m

TABLE 4.2
Characteristics of ground truth for the *Tokyo* dataset

Class number	Description
0	Water
1	Golf course
2	Park and grassland
3	Buildings
4	Forest
5	Farm

meaning that pixels encode class values, at the same spatial extent and resolution as the 10-m image. The corresponding classes are described in table 4.2. We use two distinct terrain truth sources for validation purposes: we will explain why later in section 5.2.

Download the data of the terrain truth for this chapter from the resource repository.[3] Search for the Tokyo dataset (you can use the **Find file** tool of GitHub). You should be able to download the following files:

1. terrain_truth_epsg32654_A.tif: terrain truth (split A)
2. terrain_truth_epsg32654_B.tif: terrain truth (split B)
3. legend_style.qml: the style sheet file for QGIS

Create a sub-folder of the *tokyo* directory named *terrain_truth* and download these files inside this newly created directory.

[3]`https://github.com/remicres/otbtf_tutorials_resources`

5

A simple convolutional neural network

We will start with something easy: a small convolutional neural network. First, we will look at the data that will be employed to train and use our network. Open QGIS and import the geospatial data. The raster files can be imported from the **Layer** menu as shown in figure 5.1.

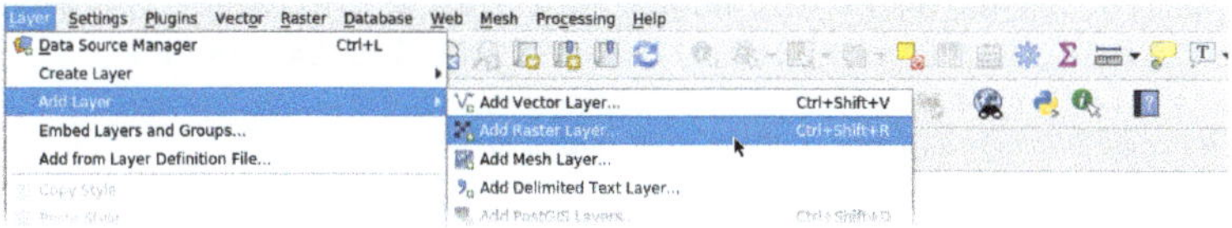

FIGURE 5.1
Open raster file.

Prepare a directory for the results somewhere on your file system, where you have the write permission accorded. We encourage the reader to use this directory to store all resulting files, and to run all command lines from inside this directory.

> **Use your favorite interface**
> The document shows how to run the Orfeo ToolBox applications from the Command Line Interface (CLI). However, it is possible to switch to the GUI (Graphical User Interface): to do so, just type *"otbgui_ApplicationName"* instead of *"otbcli_ApplicationName"*. You can also use the OTB GUI launcher (Mapla) which is located in the OTB installation directory. If you run the OTBTF from its Docker image, you might read the instructions on the OTBTF GitHub page to run Docker with the proper options to enable the graphical mode.

5.1 Normalization

In section 1, we have explained how weights of a deep net are updated during training: the training process consists of moving, little by little, the weights of the net at each iteration. Thus, distribution of the initial model weights and data are crucial to make the learning converge. In many machine learning frameworks, parameters are often defaulted to be suited for centered reduced distributions and values within the $[-1, +1]$ range. That is why data normalization is a fundamental pre-processing step. Regarding our dataset, we will shift-scale all remote sensing images channels to normalize them with their maximum value using the **BandMathX** application.

To generate the normalized image, go inside the output folder from the command line. You can use the *cd* command to navigate in the file system, which is common for Windows and UNIX.

```
# Normalize images

# Go in the folder Sentinel-2 images (10m)
cd ~/tokyo/
cd S2A_MSIL2A_20190508T012701_N0212_R074_T54SUE_20190508T041235.SAFE
cd GRANULE/L2A_T54SUE_A020234_20190508T012659/IMG_DATA/R10m/

# Concatenate images, and normalise them between [0,1]
otbcli_BandMathX \
-il T54SUE_20190508T012701_B04_10m.jp2 \
    T54SUE_20190508T012701_B03_10m.jp2 \
    T54SUE_20190508T012701_B02_10m.jp2 \
    T54SUE_20190508T012701_B08_10m.jp2 \
-exp "{im1b1/im1b1Maxi;im2b1/im2b1Maxi;im3b1/im3b1Maxi;im4b1/
im4b1Maxi}" \
-out ~/tokyo/Sentinel-2_B4328_10m.tif

# Go back in the working directory
cd ~/tokyo/
```

The command line above runs the **BandMathX** application with the following parameters:

- **il** is the input images list. We have fed 4 string values to this parameter which are the Sentinel-2 band filenames.
- **exp** is the mathematical expression that the application will perform on the pixels. The provided expression simply normalizes each channel with its maximum value, ensuring that the output range is $[0, 1]$.
- **out** is the output generated image filename.

Using the Graphical User Interface rather than the Command Line Interface

In all the following exercises, OTB applications will be run using the same *key/value* mechanism. For the sake of readability, the book provides the command lines for each exercise. However, recall that the reader can use any OTB application in graphical mode: instead of providing the parameter value with the *key/value* mechanism, the reader can perform the exercises using the graphical user interface. To run any OTB application in the graphical user interface, the user just has to change the prefix of the executable before running it from: *otbgui* (like "Graphical User Interface") instead of *otbcli* (for "Command Line Interface"). For instance, if the user wants to run the **BandMathX** application in graphical mode, he or she can directly run *otbgui_BandMathX* without additional arguments. The user interface of the **BandMathX** application will be displayed.

5.2 Sampling

The first step to apply deep learning techniques to real-world datasets is sampling. The existing framework of the Orfeo ToolBox offers great tools for pixel-wise or object-oriented classification/regression tasks. On the deep learning side, nets like CNNs are trained over patches of images rather than single-pixel batches. Hence the first application of OTBTF we will present targets patch sampling and is called **PatchesExtraction**.

5.2.1 Selection

The **PatchesExtraction** application integrates seamlessly in the existing sampling framework of OTB. The first step is the sample selection. It consists of selecting the locations of the center of the samples. There are currently two approaches depending on the type of the terrain truth format:

- When terrain truth is **vector data**: typically one can use the **PolygoncClassStatistics** and **SampleSelection** applications to select patch centers, then give them to the **PatchesExtraction** application.
- When terrain truth is **raster data**: one can use the **LabelImageSampleSelection** application, which is part of the OTBTF remote module.

Since our terrain truth is raster, we use the second approach. You can refer to the OTB CookBook[1] to see how to perform the first approach.

```
# Samples selection for group A
# You might need to change the path of terrain truth raster,
# "terrain_truth_epsg32654_A.tif" where it was downloaded
otbcli_LabelImageSampleSelection \
-inref ~/tokyo/terrain_truth/terrain_truth_epsg32654_A.tif \
-nodata 255 \
-outvec terrain_truth_epsg32654_A_pos.shp \
-strategy "constant" \
-strategy.constant.nb 500
```

[1]https://www.orfeo-toolbox.org/CookBook/

Where:

- **inref** is the input label image of the terrain truth.
- **outvec** is the output vector data of points for the physical position of samples.
- **strategy** is the sampling strategy: "constant" mode enables the sampling of an identical number of samples in each class. This number is set using the **strategy.constant.nb** parameter.

Open the generated file *terrain_truth_epsg32654_A_pos.shp* in QGIS to check the generated vector data.

> **Terrain truth split**
> For validation purposes, the terrain truth (*terrain_truth_epsg32654_A.tif* and *terrain_truth_epsg32654_B.tif*) has been split in two datasets so that the objects of *A* do not overlap objects of *B*. This way, we have two groups of objects that are completely distinct. When we perform the sample selection, this guarantees that there is no selected position from *A* and *B* that come from the same object.

Repeat the previously described steps for the second terrain truth data, *terrain_truth_epsg32654_B.tif*:

```
# Samples selection for group B
# You might need to change the path of terrain truth raster,
# "terrain_truth_epsg32654_B.tif" where it was downloaded
otbcli_LabelImageSampleSelection \
-inref ~/tokyo/terrain_truth/terrain_truth_epsg32654_B.tif \
-nodata 255 \
-outvec terrain_truth_epsg32654_B_pos.shp \
-strategy "constant" \
-strategy.constant.nb 500
```

Once sample positions are computed, open QGIS and import the dataset to verify that everything has been done correctly.

5.2.2 Extraction

Now we should have two vector layers:

- *terrain_truth_epsg32654_A_pos.shp*
- *terrain_truth_epsg32654_B_pos.shp*

It's time to use the **PatchesExtraction** application. The following operation consists of extracting patches in the input image, at each location of the *terrain_truth_epsg32654_A_pos.shp*. In order to train a small CNN later, we will create a set of patches of dimension 16×16 associated with the corresponding label given from the *class* field of the vector data. Let's do this:

```
# Patches extraction
otbcli_PatchesExtraction \
-source1.il Sentinel-2_B4328_10m.tif \
-source1.patchsizex 16 \
-source1.patchsizey 16 \
-source1.out Sentinel-2_B4328_10m_patches_A.tif \
-vec terrain_truth_epsg32654_A_pos.shp \
-field "class" \
-outlabels Sentinel-2_B4328_10m_labels_A.tif uint8
```

where:

- **source1** is the parameter group for the first image source (it cannot be set with value, and only contains parameters),
- **source1.il** is the input image list of the first source,
- **source1.patchsizex** is the patch width of the first source,
- **source1.patchsizey** is the patch height of the first source,
- **source1.out** is the output patches image of the first source,
- **vec** is the input vector data of the points (samples locations),
- **field** is the attribute that we will use as the label value (i.e. the class),
- **outlabels** is the output image for the labels.

After this step, you should have generated the following output images, that we will name "the training dataset":

- *Sentinel-2_B4328_10m_patches_A.tif*
- *Sentinel-2_B4328_10m_labels_A.tif*

Visualizing the patches
A simple trick enables you to check the extracted patches from QGIS: Simply open the patches as a raster, then chose the same coordinate reference system as the current project (indicated on the bottom right of the window).

Repeat the previously described steps for the vector layer *terrain_truth_epsg32654_B_pos.shp* and generate the patches and labels for validation.

```
# Patches extraction
otbcli_PatchesExtraction \
-source1.il Sentinel-2_B4328_10m.tif \
-source1.patchsizex 16 \
-source1.patchsizey 16 \
-source1.out Sentinel-2_B4328_10m_patches_B.tif \
-vec terrain_truth_epsg32654_B_pos.shp \
-field "class" \
-outlabels Sentinel-2_B4328_10m_labels_B.tif uint8
```

After this step, you should have the following data that will be referred as *the validation dataset*:

- *Sentinel-2_B4328_10m_patches_B.tif*
- *Sentinel-2_B4328_10m_labels_B.tif*

> **A note on pixel interleave**
> To keep things simple, sampled patches are stored in one single big image that stacks all patches in rows. There are multiple advantages to this. First, accessing one unique big file is more efficient than working on thousands of separate small files stored in the file system, or accessing sparsely located patches in large images. Secondly, we can visualize/process patches as any image: for instance, we can import the big patch image in QGIS and check that our patches look good. Thirdly, the interleave of the sampled source is preserved, which guarantees good performance during data access.

5.3 Training

How do you train a deep net? To begin with something easy, we will train a small existing model. We focus on a CNN that inputs our $16 \times 16 \times 4$ patches and produces the prediction of an output class among 6 labels ranging from 0 to 5 (see table 4.2 for class descriptions).

5.3.1 Principle

In this section, we supply a short overview of the basic concepts for the training of deep learning models. Model training usually involves a gradient descent (in the network weights space) of a loss function that typically expresses the gap between estimated and reference data. The **batch** size defines the number of samples that will be propagated through the network to average the gradient at each step. Suppose that we have N training samples and we want to use a batch size of n. During learning, the first n samples (from 1 to n) will be used to train the network. Then, the second n samples (from $n+1$ to $2n$) will be used to train the network again. This procedure is repeated until all samples are propagated through the network (this is called one **epoch**). At each step, the gradient is computed over the batch, and the network weights are updated accordingly to the chosen optimization method.

Using a batch size that is smaller than the number of samples enables the overall training procedure to require less memory, which is useful for a big dataset. Also, the network weights are updated after each propagation.

Depending on the problem, performances can be superior with multiple batches: if all samples were used during propagation, it would make only one single update for the network's parameter. However, the smaller the batch, the less accurate the estimate of the gradient will be. For each batch, the network weights will be updated to minimize the cost function that corresponds to the samples used. If the number of samples is too small, the gradient changes its direction very often compared to a bigger batch. There is no simple rule for choosing a good batch size, since this depends on the network architecture, loss functions to minimize, training parameters, data distribution, etc.

5.3.2 Model architecture

We propose to introduce a small and simple CNN model to better understand the approach. This section describes what is inside this model. An overview is provided in figure 5.2, which summarizes the computational graph of the considered architecture.

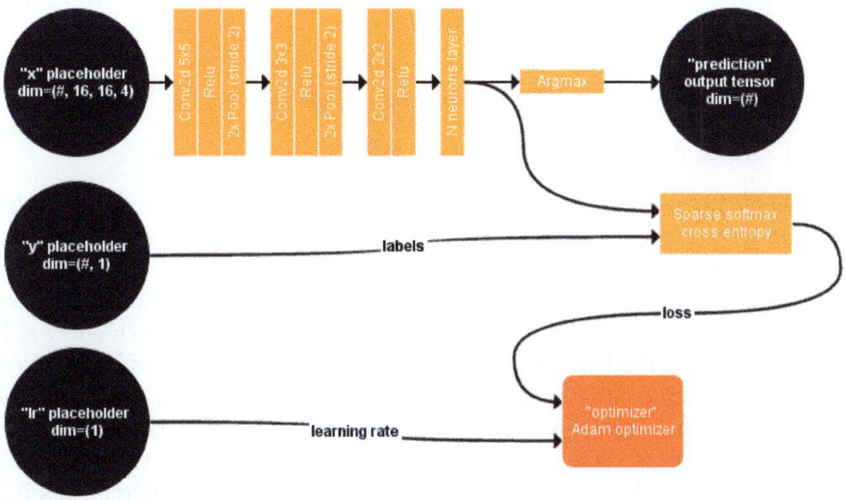

FIGURE 5.2
Our first deep net. The network consists of two placeholders (***"x"*** and ***"lr"***), respectively used for input patches (4-dimensional array) and learning rate (single scalar), one output tensor (***"prediction"***, one-dimensional array) and one target node (***"optimizer"***, used only for training the net). "#" means that the number of components for the first dimension is not fixed.

5.3.2.1 Input

The image patches feed the placeholder named ***"x"*** in the TensorFlow model. The first dimension of ***"x"*** might have any number of components. This dimension is usually employed for the batch size, and isn't fixed to enable the use of different batch sizes. For instance, assuming we want to train our model with a batch of 10 samples, we will feed the model a multidimensional array of size $10 \times 16 \times 16 \times 4$ in the placeholder ***"x"*** of size $\# \times 16 \times 16 \times 4$.

5.3.2.2 Layers

"x" is then processed by a succession of 2D-Convolution/Activation function (Rectified linear unit)/Pooling layers. After the last activation function, the features are processed by a fully connected layer of 8 neurons (one for each predicted class).

5.3.2.3 Estimated class

The estimated class is the index of the neuron (from the last neuron layer) that output the maximum value. This is performed in processing the outputs of the last fully connected layer with the *Argmax* operator, which is named ***"prediction"*** in the graph.

5.3.2.4 Loss function

The goal of the training is to minimize the cross entropy between the data distribution (real labels) and the model distribution (estimated labels). We use the cross entropy of the softmax function of the 6 neurons as the loss function. Basically, this function will measure the probability error in the discrete classification task in which the classes are mutually exclusive (each entry is in exactly one class). For this, the model implements an operation of TensorFlow known as *softmax cross entropy with logits*. This function first computes the *softmax* function of the 6-neuron outputs. The softmax function normalizes the outputs, such as their sum is equal to 1 and can be used to represent a categorical distribution, i.e, a probability distribution over n different possible outcomes (Equation 5.1). The Shannon cross entropy between true labels and probability-like values from the softmax is then computed (Equation 5.2), and considered as the loss function of the deep net.

$$\sigma(z)_i = \frac{e^{z_i}}{\sum\limits_{k=1}^{n} e^{z_k}} \tag{5.1}$$

$$H(p,q) = -\sum_{x} p(x) log(q(x)) \tag{5.2}$$

5.3.2.5 Optimizer

A node called ***"optimizer"*** performs the gradient descent of the loss function: this node is used only to train (or fine tune, if one want to "freeze" certain weights) the model. This node is not used at inference time, that is, when the model is used (typically, to estimate the classes). The method implemented in this operator is called Adam (like Adaptive moment estimation [19]). A placeholder named ***"lr"*** controls the learning rate of the optimizer: it holds a single scalar value (floating point) and can have a default value.

> **Important note**
> The functions used in our model work on labels that are integers ranging from 0 to 5 ($N = 6$). The first class number must be 0 to enforce the class numbering convention of the TensorFlow functions used in our model.

5.4 Generate the model

Now that we understand what is inside this model, we will build it. We construct a TensorFlow model using the Python API of TensorFlow. We will use the *create_model1.py* Python script provided in the resources repository.[2] You can download the code from the link, or clone the repository with Git. Before running the *create_model1.py* code with Python, the *PYTHONPATH* environment variable, which enables Python to import all the external libraries, must be properly set. The *tricks.py* Python code is part of OTBTF and contains a set of useful functions and helpers. In particular, the function *CreateSavedModel* enables us to serialize the model into a ***SavedModel*** on the file system. In the OTBTF Docker image, *tricks.py* is located in the */work/otb/otb/-Modules/Remote/otbtf/Python/* directory. We must add this directory to the *PYTHONPATH* environment variable, before running the *create_model1.py* code with Python. The *create_model1.py* code is shown and explained below.

```
#
# Do not forget to add the otbtf/python directory to your PYTHONPATH.
# Else "tricks" can't be imported.
# In the OTBTF Docker image, tricks.py is located in the following
# path: /work/otb/otb/Modules/otbtf
#
# Works with OTBTF 1.7
#
from tricks import *
import sys
```

[2]https://github.com/remicres/otbtf_tutorials_resources

```
nclasses=6

def myModel(x):

  # input patches: 16x16x4
  conv1 = tf.layers.conv2d(inputs=x, filters=16, kernel_size=[5,5],
                           padding="valid", activation=tf.nn.relu)
  # conv1 out size: 12x12x16
  pool1 = tf.layers.max_pooling2d(inputs=conv1, pool_size=[2, 2],
                                  strides=2)
  # pool1 out: 6x6x16
  conv2 = tf.layers.conv2d(inputs=pool1, filters=16, kernel_size=[3,3],
                           padding="valid", activation=tf.nn.relu)
  # conv2 out size: 4x4x16
  pool2 = tf.layers.max_pooling2d(inputs=conv2, pool_size=[2, 2],
                                  strides=2)
  # pool2 out: 2x2x16
  conv3 = tf.layers.conv2d(inputs=pool2, filters=32, kernel_size=[2,2],
                           padding="valid",activation=tf.nn.relu)
  # conv3 out size: 1x1x32

  # Features
  features = tf.reshape(conv3, shape=[-1, 32], name="features")

  # Neurons for classes
  estimated = tf.layers.dense(inputs=features, units=nclasses,
     activation=None)
  estimated_label = tf.argmax(estimated, 1, name="prediction")

  return estimated, estimated_label

""" Main """
if len(sys.argv) != 2:
  print("Usage␣:␣<output␣directory␣for␣SavedModel>")
  sys.exit(1)

# Create the TensorFlow graph
with tf.Graph().as_default():

  # Placeholders
  x = tf.placeholder(tf.float32, [None, None, None, 4], name="x")
  y = tf.placeholder(tf.int32 , [None, None, None, 1], name="y")
  default_lr = tf.constant(0.0002, dtype=tf.float32, shape=[])
  lr = tf.placeholder_with_default(default_lr, shape=[], name="lr")

  # Output
  y_estimated, y_label = myModel(x)

  # Loss function
  cost = tf.losses.sparse_softmax_cross_entropy(labels=tf.reshape
        (y, [-1, 1]), logits=tf.reshape(y_estimated,
        [-1, nclasses]))
```

```
# Optimizer
optimizer = tf.train.AdamOptimizer(learning_rate=lr,
            name="optimizer").minimize(cost)

# Initializer, saver, session
init = tf.global_variables_initializer()
saver = tf.train.Saver( max_to_keep=20 )
sess = tf.Session()
sess.run(init)

# Create a SavedModel
# Create a SavedModel
CreateSavedModel(sess=sess,
                 inputs=["x:0", "y:0"],
                 outputs=["features:0", "prediction:0"],
                 directory=sys.argv[1])
```

Important note
The Python syntax is sensitive to indentation: you must preserve the same indentation for each line under a block (e.g. function). Be careful of indentations if you copy/paste or modify the provided codes. We recommend cloning the GitHub repository instead of using copy/paste.

To run the Python script, just type:

```
# Add otbtf python directory to PYTHONPATH
export PYTHONPATH=$PYTHONPATH:/work/otb/otb/Modules/Remote/otbtf/python/

# Generate the model (you might need to provide the correct path
# to the "create_model1.py" file if you have downloaded elsewere)
# The "create_model1.py" file can be downloaded from:
# https://github.com/remicres/otbtf_tutorials_resources.git
python3 create_model1.py "model1"
```

Of course, you can change the model directory (“*model1*”) if you want to write it elsewhere than the current directory.

5.5 Train the model from scratch

If you take a look at the ***SavedModel*** directory, you will see a *.pb* file (protobuf) and a *Variables* folder. The protobuf file serializes the computational graph, and the *Variables* folder contains the values of the model weights (kernels, etc.). As you could have noticed in the Python script, the model weights are initialized before exporting the ***SavedModel*** . We will use the **TensorflowModelTrain** application to train the CNN from its initialized state, updating its weights for the image classification task. For each dataset (training data and validation data), the validation step of the **Tensorflow-ModelTrain** application consists of computing the following metrics:

- Confusion matrix
- Overall accuracy index
- Kappa index
- For each class:
  - Precision
  - Recall
  - F-Score

Depending on the placeholder names of the sources used for training, and the names of the sources used for evaluation, the application to detect which source is used as input of the model, or as a baseline to compute evaluation metrics, is explained in figure 5.3.

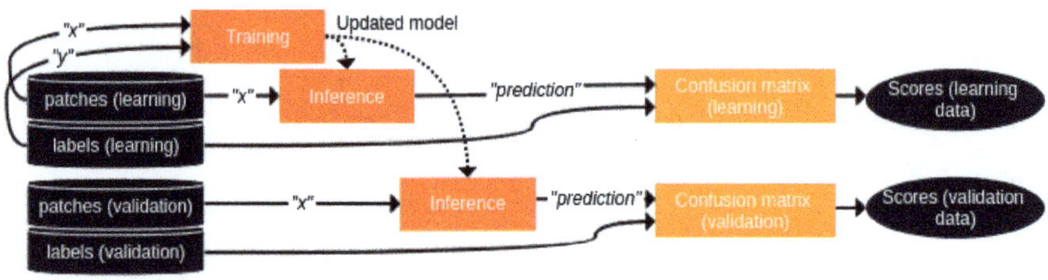

FIGURE 5.3
Computation of classification metrics, performed separately on the training dataset and validation dataset.

The application parameters are the following:

- **training.source1** is the parameter group for the first source, i.e. image patches (for learning)
  - **training.source1.il** is the input image filename
  - **training.source1.patchsizex** is the patch width
  - **training.source1.patchsizey** is the patch height
  - **training.source1.placeholder** is the name of the placeholder for the image patches
- **training.source2** is the parameter group for the second source, i.e. label patches (for learning)

  - **training.source2.il** is the input image filename
  - **training.source2.patchsizex** is the label width
  - **training.source2.patchsizey** is the label height
  - **training.source2.placeholder** is the name of the placeholder for the label patches

- **model.dir** is the directory containing the TensorFlow ***SavedModel***
- **training.targetnodes** is the name of the operator that computes the gradient descent in the training step. In our model, the gradient descent is performed with the adam optimizer called ***"optimizer."***
- **validation.mode** is the validation mode. The "**class**" validation mode enables the computation of classification metrics for both training data and validation data.
- **validation.source1** is the patch source (for validation). As the name of the source for validation (**validation.source1.name**) is the same as the placeholder name of the same source for training (**training .source1.placeholder**), this source is considered as an input of the model, and is fed to the corresponding placeholder during the validation step.
- **validation.source2** is the label source (for validation). As the name of the source (**validation. source2.name**) is different than the placeholder name of the same source for training (**training.source2.placeholder**), this source is considered as a reference to be compared to the output of the model that has the same tensor name during the validation step.
- **model.saveto** enables us to export the model variables (i.e. weights) to a file.

The command line corresponding to the above description is the following:

```
# Train the deep learning model
otbcli_TensorflowModelTrain \
-training.source1.il Sentinel-2_B4328_10m_patches_A.tif \
-training.source1.patchsizex 16 \
-training.source1.patchsizey 16 \
-training.source1.placeholder "x" \
-training.source2.il Sentinel-2_B4328_10m_labels_A.tif \
-training.source2.patchsizex 1 \
-training.source2.patchsizey 1 \
-training.source2.placeholder "y" \
-model.dir model1 \
-training.targetnodes "optimizer" \
-validation.mode "class" \
-validation.source1.il Sentinel-2_B4328_10m_patches_B.tif \
-validation.source1.name "x" \
-validation.source2.il Sentinel-2_B4328_10m_labels_B.tif \
-validation.source2.name "prediction" \
-model.saveto model1/variables/variables
```

As you can notice, the *training.source1.placeholder* and *validation.source1 .name* are the same (*"x"*), and *training.source2.placeholder* and *validation.source2.name* are different (*"y"* versus ***"prediction"***). This causes the placeholder ***x*** to be used for both training and validation, but the placeholder ***y*** is used only for training, and the placeholder ***prediction*** for validation, as explained in figure 5.3.

We could tune the learning rate (which has, as mentioned in the Python code, a default value of 0.0002) using the *training.userplaceholders* parameter:

```
...
-training.userplaceholders "lr=0.0005" \
...
```

Run the **TensorflowModelTrain** application. After the epochs, note the kappa and overall accuracy indexes (should be, respectively, around 0.7 over the validation dataset). Browse the file system, and take a look in the *data/results/model1/variables/* directory: you can notice that the application has updated two files:

- ***model_variables_cnn.index*** is a summary of the saved variables,
- ***model_variables_cnn.data-00000-of-00001*** is the saved variable number 1, saved at the end of the training process.

5.6 Comparison with Random Forest

Here, we compare our small deep net to a Random Forest (RF) classifier. We use the metrics deriving from the confusion matrix. We use the machine learning framework of OTB to train and evaluate the performance of the RF classifier. First, we use exactly the same samples as for the deep net training and validation, to extract the pixel values of the image:

```
# Extract the pixels values at samples locations (Learning data)
otbcli_SampleExtraction \
-in Sentinel-2_B4328_10m.tif \
-vec terrain_truth_epsg32654_A_pos.shp \
-field "class" \
-out terrain_truth_epsg32654_A_pixelvalues.shp
```

Then, we do the same with the *terrain_truth_epsg32654_B_pos.shp* vector data (validation data).

```
# Extract the pixels values at samples locations (Validation data)
otbcli_SampleExtraction \
-in Sentinel-2_B4328_10m.tif \
-vec terrain_truth_epsg32654_B_pos.shp \
-field "class" \
-out terrain_truth_epsg32654_B_pixelvalues.shp
```

Finally, we train an RF classifier with default parameters.

```
# Train a Random Forest classifier with validation
otbcli_TrainVectorClassifier \
-io.vd terrain_truth_epsg32654_A_pixelvalues.shp \
-valid.vd terrain_truth_epsg32654_B_pixelvalues.shp \
-feat "value_0" "value_1" "value_2" "value_3" \
-cfield "class" \
-classifier "rf" \
-io.out randomforest_model.yaml
```

Note the kappa index: it should be around 0.5, i.e. about 20% less than our CNN model.

> **Important note**
> The CNN uses more information for training, compared to the RF. Indeed, it learns on 16×16 patches of a multispectral image, whereas the RF learns only on multispectral pixels (that could be interpreted as a 1×1 patches of multispectral image). Our goal here is not to compare RF vs CNN, but just to show that the contextual information process by the CNN is useful in classification task. A more fair competitor could be an RF using spatial features like textures, local Fourier transforms, SIFTs, etc.

5.7 Inference

In section 5.3 we successfully trained our CNN model. Now we will apply the model over the Sentinel-2 image to produce a land cover map. The optimizer is not used anymore, we just employ the model with its fixed weights as updated after the training. This part is called *the inference.* To perform the inference, we use the **TensorflowModelServe** application. We know that our CNN input has a receptive field of 16×16 pixels and the placeholder name is ***"x."*** The output of the model is the estimated class, which is the tensor result of the *Argmax* operator, named ***"prediction."*** We don't use the optimizer node anymore, as it is part of the training procedure: we use only the model subset that computes ***"prediction"*** from ***"x"*** as shown in figure 5.4.

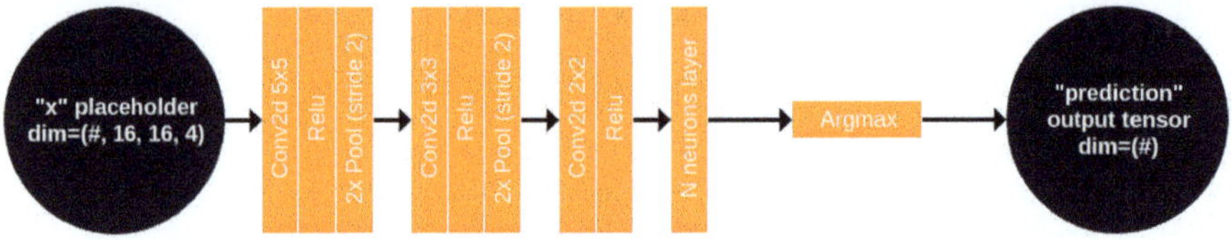

FIGURE 5.4
For the inference, we use only the placeholders (***"x"***) and we compute the one output tensor named (***"prediction,"*** one-dimensional array).

As we don't have GPU support for now, it could be slow to process the whole image. We won't produce the map over the entire image (even if that's possible thanks to the streaming mechanism) but just over a small subset. We do this using the extended filename of the output image, setting a subset starting at pixel 4000, 4000 with size 1000×1000. This extended filename consists of adding **?&box=4000:4000:1000:1000** to the output image filename. Note that you can also generate a small image subset with the **ExtractROI** application of OTB, then use it as input of the **TensorflowModelServe**.

Before running the command line, here is a quick summary of the application parameters:

- **source1** is the parameter group for the first image source,
- **source1.il** is the input image list of the first source,
- **source1.rfieldx** is the receptive field width of the first source,
- **source1.rfieldy** is the receptive field height of the first source,
- **source1.placeholder** is placeholder name corresponding to the first source in the TensorFlow model,
- **model.dir** is the directory of the***SavedModel***,
- **output.names** is the list of the output tensors that will be produced then generated as the output image,
- **out** is the filename for the output image generated from the TensorFlow model applied to the entire input image.

```
otbcli_TensorflowModelServe \
-source1.il Sentinel-2_B4328_10m.tif \
-source1.rfieldx 16 \
-source1.rfieldy 16 \
-source1.placeholder "x" \
-model.dir model1 \
-output.names "prediction" \
-out "classif_model1.tif?&box=4000:4000:1000:1000"
```

Now import the generated image in QGIS ref to figure 5.5. You can change the style of the raster: in the **layers panel** (left side of the window), right-click on the image, then select **Properties**, go to the **Symbology** tab and select **render type** as **single band pseudocolor**. Then, you can select the color for each class value, the annotations, etc. You can also load the predefined style by clicking on the **Style** button (bottom of the window) and opening the file named *legend_style.qml* stored in the *tokyo/terrain_truth* directory.

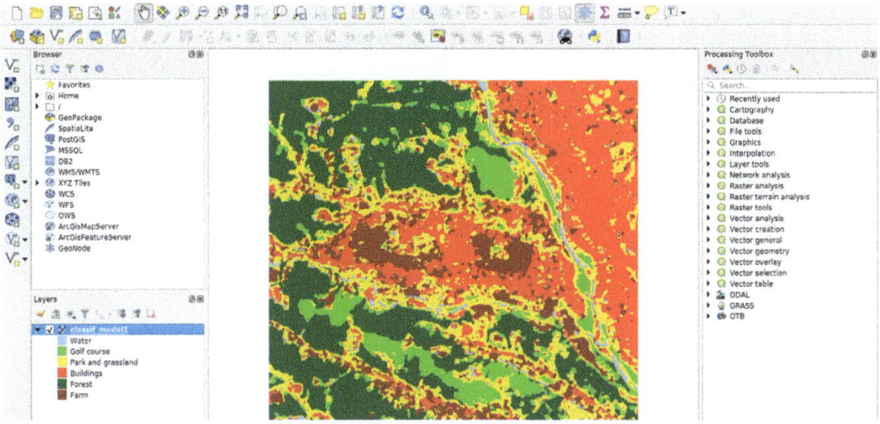

FIGURE 5.5
Import the output classification map in QGIS.

> **Important note**
> We just have run the CNN in **patch-based** mode, meaning that the application extracts and process patches independently at regular intervals. This is costly, because the sampling strategy requires that we duplicate a lot of overlapping patches and process them independently.

6

Fully Convolutional Neural Network

In the previous section, we performed pixel-wise classification using a deep convolutional neural network. We have performed the inference in patch-based mode, meaning that for each output pixel, we have run the model on one small patch of the input image, centered on the output pixel position. While this kind of network architecture is easy to implement, it is not efficient in terms of processing. Due to patch overlap, the data is copied multiple times with different memory alignment (for each different patch) and passed to the model. The costliest operations implemented in the deep net are the convolutions, which are massively parallel and can be implemented so that the intermediary results are reused for other patch processing (e.g. intermediate sums, see equations 1.2 and 1.3). However, the patch-based mode disables the possible reuse of the intermediate results since they are re-computed at each new patch, which is not efficient. We will introduce the Fully Convolutional Neural Networks (FCNN, or FCN) that are just CNNs that can process entire image regions instead of being limited to small patches.

6.1 Using the existing model as an FCN

From the constitution of the simple CNN we used in section 5.4, we can notice that this model can be used as an FCN. Its properties are summarized as the following:

- **receptive field** (16×16 pixels),
- **expression field** (1×1 pixel),
- **scale factor** (there is a total of 2 successive pooling operators of size 2, meaning that the scale factor is $2 \times 2 = 4$).

We can check that they are consistent with operators implemented in the model. If we let n be the size of the input image, we can see how the model propagates the image region to the output of size m:

1. convolution with 5×5 kernel: output size is $n - 4$,
2. 2x pooling with stride $(2, 2)$: output size is $(n - 4)/2 = n/2 - 2$,

3. convolution with 3×3 kernel: output size is $(n/2-2)-2 = n/2-4$,
4. 2x pooling with stride $(2,2)$: output size is $(n/2-4)/2 = n/4-2$,
5. convolution with 2×2 kernel: output size is $(n/4-2)-1 = n/4-3$.

In conclusion, we have $m = n/4 - 3$, which is quite consistent with the properties (receptive field, expression field, spacing factor) of the model: the model produces an output of size 1×1 from an input image of size 16×16. The physical spacing of its output pixel is equal to the input pixel spacing divided by 4. Run the CNN model with the parameter **model.fullyconv** to **on** in order to enable the fully convolutional processing.

```
otbcli_TensorflowModelServe \
-source1.il Sentinel-2_B4328_10m.tif \
-source1.rfieldx 16 \
-source1.rfieldy 16 \
-source1.placeholder "x" \
-model.dir model1 \
-model.fullyconv on \
-output.names "prediction" \
-output.spcscale 4 \
-out "classif_model1_fcn.tif?&box=1000:1000:250:250"
```

Open the resulting image in QGIS, and compare with the classification map created by the original CNN. Open the **Properties** > **Metadata** tab and check that the physical spacing of the new map is 4 times greater than the original.

6.2 Pixel-wise fully convolutional model

A fully convolutional network runs faster than patch based, enough to process the full image in reasonable time (to test it, just remove the extended filename in the previous command). However, pooling completely destroys the resolution of the output (the output image spacing is 4 times greater than the input pixel spacing). We can modify our original model a bit to create a pixel-wise fully convolutional network which preserves the input image spacing.

We remove the pooling operators, and add some convolutional layers with unit strides. Our goal is to keep the same receptive field and expression field as the previously used architecture (this way, we can reuse our image patches that we have generated in section 5.2). Figure 6.1 summarizes our new architecture.

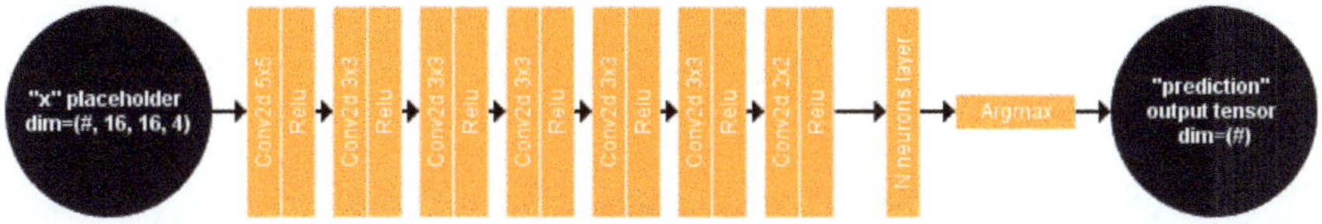

FIGURE 6.1
The new Fully Convolutional Neural Network. Pooling operators have been removed, and convolution operators have been added. Note that there is no stride in convolution, preserving the physical spacing in spatial dimensions.

Let's implement our new FCN architecture. We will use the *create_model2.py* Python code provided in the resources repository.[1]

```
#
# Do not forget to add the otbtf/python directory to your PYTHONPATH.
# Else "tricks" can't be imported.
# In the OTBTF Docker image, tricks.py is located in the following
# path: /work/otb/otb/Modules/otbtf
#
# Works with OTBTF 1.7
#
from tricks import *
import sys

nclasses=6

def myModel(x):

  # input patches: 16x16x4
  conv1 = tf.layers.conv2d(inputs=x, filters=16, kernel_size=[5,5],
                           padding="valid", activation=tf.nn.relu)
  # conv1 out size: 12x12x16
  conv2 = tf.layers.conv2d(inputs=conv1, filters=16, kernel_size=[3,3],
                           padding="valid", activation=tf.nn.relu)
  # conv2 out size: 10x10x16
  conv3 = tf.layers.conv2d(inputs=conv2, filters=16, kernel_size=[3,3],
                           padding="valid", activation=tf.nn.relu)
  # conv3 out size: 8x8x16
  conv4 = tf.layers.conv2d(inputs=conv3, filters=32, kernel_size=[3,3],
                           padding="valid", activation=tf.nn.relu)
  # conv4 out size: 6x6x32
  conv5 = tf.layers.conv2d(inputs=conv4, filters=32, kernel_size=[3,3],
                           padding="valid", activation=tf.nn.relu)
  # conv5 out size: 4x4x32
  conv6 = tf.layers.conv2d(inputs=conv5, filters=32, kernel_size=[3,3],
                           padding="valid", activation=tf.nn.relu)
  # conv6 out size: 2x2x32
```

[1] https://github.com/remicres/otbtf_tutorials_resources

```
  conv7 = tf.layers.conv2d(inputs=conv6, filters=32, kernel_size=[2,2],
                           padding="valid", activation=tf.nn.relu)
  # conv7 out size: 1x1x32

  # Features
  features = tf.reshape(conv7, shape=[-1, 32], name="features")

  # Neurons for classes
  estimated = tf.layers.dense(inputs=features, units=nclasses,
      activation=None)
  estimated_label = tf.argmax(estimated, 1, name="prediction")

  return estimated, estimated_label

""" Main """
if len(sys.argv) != 2:
  print("Usage␣:␣<output␣directory␣for␣SavedModel>")
  sys.exit(1)

# Create the TensorFlow graph
with tf.Graph().as_default():

  # Placeholders
  x = tf.placeholder(tf.float32, [None, None, None, 4], name="x")
  y = tf.placeholder(tf.int32 , [None, None, None, 1], name="y")
  default_lr = tf.constant(0.0002, dtype=tf.float32, shape=[])
  lr = tf.placeholder_with_default(default_lr, shape=[], name="lr")

  # Output
  y_estimated, y_label = myModel(x)

  # Loss function
  cost = tf.losses.sparse_softmax_cross_entropy(labels=tf.reshape
          (y, [-1, 1]), logits=tf.reshape(y_estimated, [-1, nclasses]))

  # Optimizer
  optimizer = tf.train.AdamOptimizer(learning_rate=lr,
              name="optimizer").minimize(cost)

  # Initializer, saver, session
  init = tf.global_variables_initializer()
  saver = tf.train.Saver( max_to_keep=20 )
  sess = tf.Session()
  sess.run(init)

  # Create a SavedModel
  CreateSavedModel(sess=sess,
                   inputs=["x:0", "y:0"],
                   outputs=["features:0", "prediction:0"],
                   directory=sys.argv[1])
```

Run the *create_model2.py* Python script from the terminal to generate the ***SavedModel*** of our new FCN architecture:

```
# Generate the model

# The "create_model2.py" file can be downloaded from:
# https://github.com/remicres/otbtf_tutorials_resources.git
python3 ~/python/create_model2.py "model2"
```

As there is no stride in convolution, and stride in pooling, our new model has an output scale factor of 1 and can produce a classification map at the same scale as the input image!

6.3 Training

We can now train this new model over the same patches as for the simple CNN.

> **Important note**
> The proposed FCN has a lot more weight than the previously introduced CNN one. We can then expect that the gradient descent will take more time. It could be better to set a higher number of epochs!

To train the new model, just run:

```
# Train the deep learning model
otbcli_TensorflowModelTrain \
-training.source1.il Sentinel-2_B4328_10m_patches_A.tif \
-training.source1.patchsizex 16 \
-training.source1.patchsizey 16 \
-training.source1.placeholder "x" \
-training.source2.il Sentinel-2_B4328_10m_labels_A.tif \
-training.source2.patchsizex 1 \
-training.source2.patchsizey 1 \
-training.source2.placeholder "y" \
-model.dir model2 \
-training.targetnodes "optimizer" \
-validation.mode "class" \
-validation.source1.il Sentinel-2_B4328_10m_patches_B.tif \
-validation.source1.name "x" \
-validation.source2.il Sentinel-2_B4328_10m_labels_B.tif \
-validation.source2.name "prediction" \
-model.saveto model2/variables/variables
```

After 100 epochs, we can notice that the scores are great and that this FCN model outperforms the simple CNN. That might come from the fact that we have significantly increased the number of variables in the model (compared to our first CNN, we added 2 new convolution operators, and we also have enlarged all convolution kernels of the network).

6.4 Inference

We now want to produce the classification map over the entire image. We run the model over the image subset:

```
otbcli_TensorflowModelServe \
-source1.il Sentinel-2_B4328_10m.tif \
-source1.rfieldx 16 \
-source1.rfieldy 16 \
-source1.placeholder "x" \
-model.dir model2 \
-model.fullyconv on \
-output.names "prediction" \
-out "classif_model2.tif?&box=4000:4000:1000:1000"
```

The map can be produced over the whole image. The **optim** parameter group of **TensorflowModelServe** enables us to adjust settings for the model execution to reach better performance. To speed up the process, you can force the application to produce the result using large tiles. In the following, we force the output tiles to large rectangular tiles of 128-pixel height, in order to reduce the computation from partially overlapping input areas.

```
# Generate the landcover map over the entire image
otbcli_TensorflowModelServe \
-source1.il Sentinel-2_B4328_10m.tif \
-source1.rfieldx 16 \
-source1.rfieldy 16 \
-source1.placeholder "x" \
-model.dir model2 \
-model.fullyconv on \
-output.names "prediction" \
-optim.tilesizex 999999 \
-optim.tilesizey 128 \
-out classif_model2_full.tif uint8
```

The processing of the entire image lasts approximately **80 seconds** on an i7 Intel(R) Xeon(R) with CPU E5-1650 @ 3.20 GHz with 16 GB RAM. You can open the land cover map in QGIS.

Important note
You must have enough memory for processing large tiles. If there is not enough memory, TensorFlow will throw warning messages indicating that the efficiency of the computation is compromised. In this case, you can reduce the tile height, e.g. 64 instead of 128.

7

Classifiers on deep features

Recent studies have pointed out that using the features computed with deep nets in machine learning algorithms like Support Vector Machines or Random Forests, offers state-of-the-art results [20, 21, 22]. This section provides insight into classification using classic Machine Learning techniques on deep learning features. Here, deep learning is used merely as feature engineering rather than a sole classifier.

7.1 Principle

We will use algorithms that are already implemented in the Orfeo ToolBox with features coming from our deep net. The whole workflow is summarized in figure 7.1.

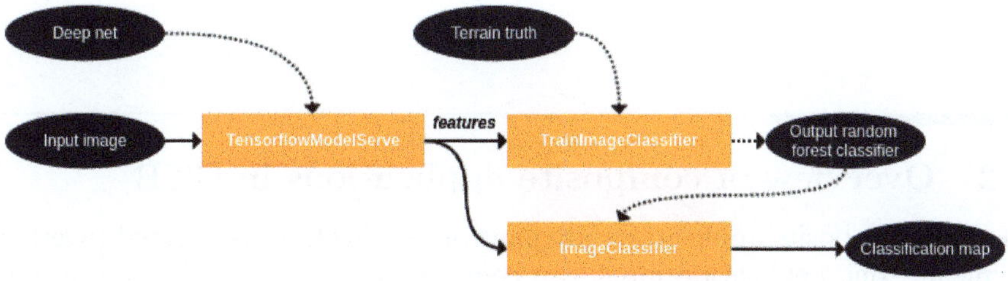

FIGURE 7.1
Workflow for hybrid classification, mixing deep nets with random forest.

In section 5.3.2, we have described the model guts. In particular, before computing the classes pseudo-probability on the last neurons, our model computes a number of features, as shown on figure 7.2. This is done by reshaping the output of the last convolution operator to a vector of dimension 2: one dimension for the batch size, which is not fixed, the other for the components of the features, which is fixed (32 elements). In a TensorFlow model, once an operator has a name, it is possible to fetch its result during the inference.

Since the *tf.reshape* operator performing this is named ***features***, it is possible to use the **TensorflowModelServe** application to produce the ***features*** for an entire image, using the same mechanisms as described in previous sessions.

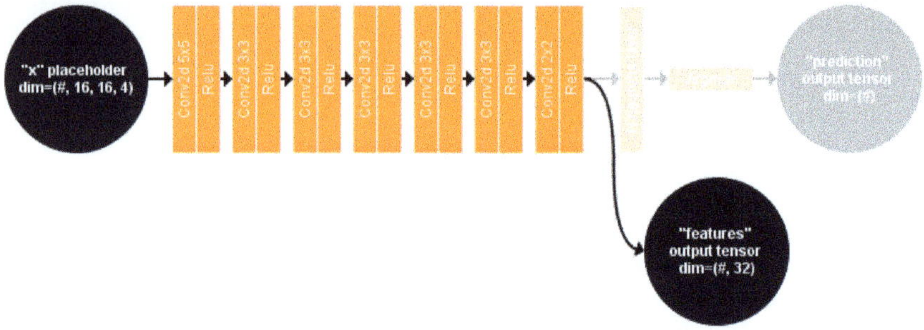

FIGURE 7.2
The model outputs the tensor named ***"features."*** The output of the dense layer composed of 6 neurons is not used. Only ***"features"*** is computed using the net, and the classification is performed using another classifier.

This enables us to use the **TensorflowModelServe** to produce the ***features*** output tensor for pixels of the image, and the resulting pixels are then used as features for training a Random Forest (with the **TrainImageClassifier** application). Once the RF rule is produced, it can be used to classify the entire image from the deep features computed on each pixel.

7.2 Overview of composite applications in OTB

In OTB, composite applications are multiple applications connected together. Typically, one application input can come from another application output, instead of being read from file. The major benefit is that temporary file read and write are avoided, which (i) saves some disk space, and (ii) helps in reducing processing time when I/O is a bottleneck. Presented in section 2.1.2, the streaming mechanism enables the processing of large images, in splitting the requested parts of the images into multiple regions. Composite applications that involve applications with streaming support, also benefit from this mechanism.

In OTBTF there are currently two composite applications.

- **TrainClassifierFromDeepFeatures**, which reuses our model serving application (**TensorflowModelServe**) as input of the *TrainImagesClassifier* to train a machine learning algorithm based on the features of the deep net given in the first application.

- **ImageClassifierFromDeepFeatures** uses **TensorflowModelServe** as input of the **ImagesClassifier** application to classify the input image.

> **Important note**
> Since the classifier is based on the features computed from the network, the deep net used in **ImageClassifierFromDeepFeatures** must be the same as the deep net used in **TrainClassifierFromDeepFeatures** to generate the classification rule! If not, the classifier would perform the image classification task using the feature vectors coming from the wrong model, and the classification would be irrelevant.

7.3 Training

This exercise will require a slightly more important amount of computing power, because the whole input image will be processed with the deep net twice. The first time: for the classifier training (using the **TrainClassifierFromDeepFeatures** application). The second time: during the classification map generation (using the **ImageClassifierFromDeepFeatures** application).

> **Important note**
> This processing lasts approximately **3 minutes** on a i7 Intel(R) Xeon(R) with CPU E5-1650 @ 3.20 GHz with 16 GB RAM. You can adjust the parameters of the processing using the ***optim*** parameter group of the applications, using the same parameter values as you did in section 6.2. You can do it for **ImageClassifierFromDeepFeatures**. However, computations of **TrainClassifierFromDeepFeatures** are mainly controlled by the ***ram*** parameter, which determines the requested region using an internal streaming mechanism based on the tile hint of the images. It is thus advised to set a small value for the ***ram*** parameter in this application, to avoid memory allocation issues.

For the training, you can also set the ***io.confmatout*** parameter to 1, to produce an output confusion matrix.

```
otbcli_TrainClassifierFromDeepFeatures \
-source1.il Sentinel-2_B4328_10m.tif \
-source1.rfieldx 16 \
-source1.rfieldy 16 \
-source1.placeholder "x" \
-model.dir model2 \
-model.fullyconv on \
-output.names "features" \
```

```
-vd terrain_truth_epsg32654_A_pos.shp \
-valid terrain_truth_epsg32654_B_pos.shp \
-sample.vfn "class" \
-sample.bm 0 \
-classifier rf \
-optim.tilesizex 999999 \
-optim.tilesizey 128 \
-out RF_model_from_deep_features.yaml
```

Once the training has been finished, you should have a *.yaml* file that serializes the random forest classification rule, in a human-readable format.

7.4 Inference

We use the **ImageClassifierFromDeepFeatures** to generate the whole classification map, using the hybrid RF deep-net approach.

```
# Generate a subset of the map using the Random Forest classifier
# performing on the features coming from the deep net
otbcli_ImageClassifierFromDeepFeatures \
-source1.il Sentinel-2_B4328_10m.tif \
-source1.rfieldx 16 \
-source1.rfieldy 16 \
-source1.placeholder "x" \
-deepmodel.dir model2 \
-deepmodel.fullyconv on \
-output.names "features" \
-model RF_model_from_deep_features.yaml \
-out "RF_model_from_deep_features_map.tif?&box=4000:4000:1000:1000"
 uint8
```

Analyze the results in checking the classification metrics, and importing the output classification map in QGIS.

8

Dealing with multiple sources

Deep nets can be used efficiently to deal with multiple heterogeneous sources, for instance, in implementing a number of branches tailored for each data source, that eventually merge together at some point to fuse the information. For example, it is easy to build a deep neural network that can deal simultaneously with time series and one very-high-resolution image: while one branch composed of a recurrent neural network processes the time series (For instance Sentinel-1 or Sentinel-2 images), another branch composed of a CNN processes the high-resolution image (for instance Spot 6/7 images). Such approaches are available in the literature (for instance in [13]).

This section will present how OTBTF can be used with any number of image sources.

8.1 More sources?

The number of sources in OTBTF applications can be controlled using the environment variable **OTB_TF_NSOURCES**. If this environment variable is undefined, the default value for the number of sources is 1. You can change it using the command line. In the following, we will use 2 sources from the Sentinel-2 image:

1. Channels at 20-m spacing (Spectral bands 5, 6, 7, 8a, 11 and 12), stacked in a single raster
2. Channels at 10-m spacing (Spectral bands 4, 3, 2 and 8), stacked in a single raster

The architecture inputs two images at different resolution to derive a single output at the desired resolution. This architecture is inspired from [23].

Let's change the number of sources from the terminal:

```
export OTB_TF_NSOURCES=2
```

Then, you can check that the number of sources has been increased to 2 in every applications of OTBTF. For instance, type the following in the terminal:

```
otbcli_PatchesExtraction -help
```

or

```
otbcli_TensorflowModelServe -help
```

As you can notice, the applications now have 2 sources, namely ***source1*** and ***source2***, each one with its parameters.

> **Important note**
> If you try with the **TensorflowModelServe** application, you can notice that there is
> **OTB_TF_NSOURCES**+1 sources. This comes from the fact that there is at least one more source for the reference input: in the case of the classification task for instance, this additional input is required for the labels of the ground truth.

8.2 Model with multiple inputs

We propose a CNN with two inputs with different resolution: the 10-m-spacing bands of Sentinel-2 (Spectral band 2, 3, 4 and 8) and the 20-m-spacing bands of Sentinel-2 (Spectral band 5, 6, 7, 8a, 11 and 12). This model processes the two inputs jointly and outputs a land cover map at the lowest resolution, i.e. 20-m spacing. The 20-m-spacing image patches have a size of 8×8 and the 10-m-spacing image a size of 16×16. Note that we could have different patch sizes, but we want our model inputs to be the same physical area in both resolutions. Figure 8.1 summarizes the implemented deep net.

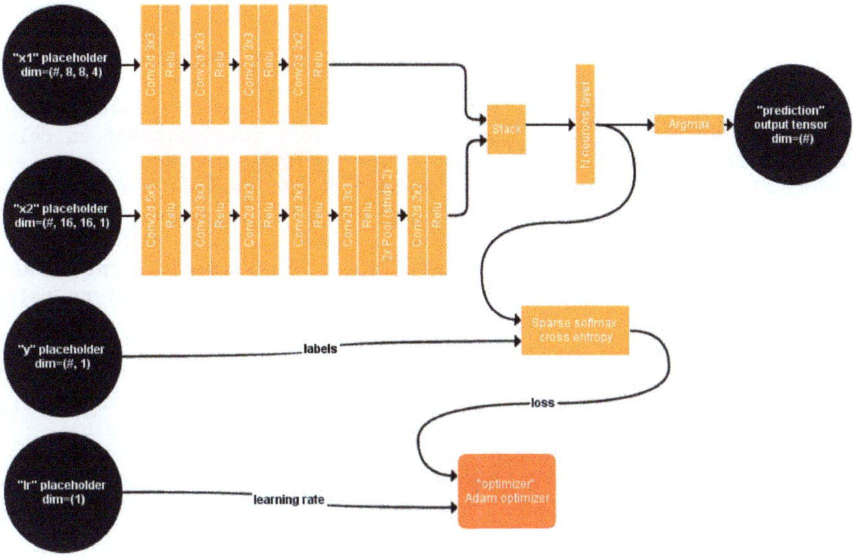

FIGURE 8.1
The network consists of three placeholders (***"x1,"*** ***"x2,"*** and ***"lr"***) respectively used for input images at 20 m (4-dimensional array), input images at 10 m (4-dimensional array), and learning rate (single scalar), one output tensor (***"prediction,"*** one-dimensional array) and one target node (***"optimizer,"*** used only for training the net).

Use the *create_model3.py* Python code provided in the resources repository[1] to generate a ***SavedModel*** that implements the double branch CNN described above.

```
#
# Do not forget to add the otbtf/python directory to your PYTHONPATH.
# Else "tricks" can't be imported.
# In the OTBTF Docker image, tricks.py is located in the following
# path: /work/otb/otb/Modules/otbtf
#
# Works with OTBTF 1.7
#
from tricks import *
import sys

nclasses=6

def myModel(x1,x2):
```

[1]https://github.com/remicres/otbtf_tutorials_resources

```
    # The 20m spacing branch (input patches: 8x8x3)
    conv1_x1 = tf.layers.conv2d(inputs=x1, filters=16, kernel_size=[3,3],
                                padding="valid", activation=tf.nn.relu)
    # conv1_x1 out size: 6x6x16
    conv2_x1 = tf.layers.conv2d(inputs=conv1_x1, filters=32, kernel_size=
                               [3,3], padding="valid", activation=tf.nn.relu)
    # conv2_x1 out size: 4x4x32
    conv3_x1 = tf.layers.conv2d(inputs=conv2_x1, filters=64, kernel_size=
                               [3,3], padding="valid", activation=tf.nn.relu)
    # conv3_x1 out size: 2x2x64
    conv4_x1 = tf.layers.conv2d(inputs=conv3_x1, filters=64, kernel_size=
                               [2,2], padding="valid", activation=tf.nn.relu)
    # conv4_x1 out size: 1x1x64

    # The 10m spacing branch (input patches: 16x16x4)
    conv1_x2 = tf.layers.conv2d(inputs=x2, filters=16, kernel_size=[5,5],
                                padding="valid", activation=tf.nn.relu)
    # conv1_x2 out size: 12x12x16
    conv2_x2 = tf.layers.conv2d(inputs=conv1_x2, filters=32, kernel_size=
                               [3,3], padding="valid", activation=tf.nn.relu)

    # conv2_x2 out size: 10x10x32
    conv3_x2 = tf.layers.conv2d(inputs=conv2_x2, filters=64, kernel_size=
                               [3,3], padding="valid", activation=tf.nn.relu)
    # conv3_x2 out size: 8x8x64
    conv4_x2 = tf.layers.conv2d(inputs=conv3_x2, filters=64, kernel_size=
                               [3,3], padding="valid", activation=tf.nn.relu)
    # conv4_x2 out size: 6x6x64
    conv5_x2 = tf.layers.conv2d(inputs=conv4_x2, filters=64, kernel_size=
                               [3,3], padding="valid", activation=tf.nn.relu)
    # conv5_x2 out size: 4x4x64
    pool1_x2 = tf.layers.max_pooling2d(inputs=conv5_x2, pool_size=[2, 2],
                                       strides=2)
    # pool1_x2 out size: 2x2x64
    conv6_x2 = tf.layers.conv2d(inputs=pool1_x2, filters=64, kernel_size=
                               [2,2], padding="valid", activation=tf.nn.relu)
    # conv6_x2 out size: 1x1x64

    # Stack features from the two branches
    features = tf.reshape(tf.stack([conv4_x1, conv6_x2], axis=3),
                          shape=[-1, 128], name="features")

    # Fully connected layer
    dense_1 = tf.layers.dense(inputs=features, units=128, activation=
                          tf.nn.relu)

    # Neurons for classes
    estimated = tf.layers.dense(inputs=dense_1, units=nclasses,
                          activation=None)
    estimated_label = tf.argmax(estimated, 1, name="prediction")

    return estimated, estimated_label

""" Main """
```

```
# check number of arguments
if len(sys.argv) != 2:
  print("Usage␣:␣<output␣directory␣for␣SavedModel>")
  sys.exit(1)

# Create the graph
with tf.Graph().as_default():

  # Placeholders
  x1 = tf.placeholder(tf.float32, [None, None, None, 6], name="x1")
  x2 = tf.placeholder(tf.float32, [None, None, None, 4], name="x2")
  y = tf.placeholder(tf.int32 , [None, None, None, 1], name="y")
  default_lr = tf.constant(0.0002, dtype=tf.float32, shape=[])
  lr = tf.placeholder_with_default(default_lr, shape=[], name="lr")

  # Output
  y_estimated, y_label = myModel(x1,x2)

  # Loss function
  cost = tf.losses.sparse_softmax_cross_entropy(labels=tf.reshape
          (y, [-1, 1]), logits=tf.reshape(y_estimated, [-1, nclasses]))
  # Optimizer
  optimizer = tf.train.AdamOptimizer(learning_rate=lr,
               name="optimizer").minimize(cost)

  # Initializer, saver, session
  init = tf.global_variables_initializer()
  saver = tf.train.Saver( max_to_keep=20 )
  sess = tf.Session()
  sess.run(init)

  # Create a SavedModel
  CreateSavedModel(sess=sess,
                 inputs=["x1:0", "x2:0", "y:0"],
                 outputs=["features:0", "prediction:0"],
                 directory=sys.argv[1])
```

You can then create the ***SavedModel*** of this new architecture from the terminal using:

```
# Generate the model

# The "create_model3.py" file can be downloaded from:
# https://github.com/remicres/otbtf_tutorials_resources.git
python3 ~/python/create_model3.py "model3"
```

8.3 Normalization

We normalize the images which are at 4-dimensional 20-m spacing using the **BandMathX** application, as explained in section 5.1.

Run the following command line:

```
# Normalize images

# Go in Sentinel-2 images (20m)
cd ~/tokyo
cd S2A_MSIL2A_20190508T012701_N0212_R074_T54SUE_20190508T041235.SAFE
cd GRANULE/L2A_T54SUE_A020234_20190508T012659/IMG_DATA/R20m

# Concatenate images, and normalise them between [0,1]
otbcli_BandMathX \
-il T54SUE_20190508T012701_B05_20m.jp2 \
    T54SUE_20190508T012701_B06_20m.jp2 \
    T54SUE_20190508T012701_B07_20m.jp2 \
    T54SUE_20190508T012701_B8A_20m.jp2 \
    T54SUE_20190508T012701_B11_20m.jp2 \
    T54SUE_20190508T012701_B12_20m.jp2 \
-exp "{im1b1/im1b1Maxi;im2b1/im2b1Maxi;im3b1/im3b1Maxi;␣\
im4b1/im4b1Maxi;im5b1/im5b1Maxi;im6b1/im6b1Maxi}" \
-out ~/tokyo/Sentinel-2_B567_20m.tif

# Go back in the working directory
cd ~/tokyo
```

8.4 Sampling

Now we will use the **PatchesExtraction** application to create two patch images, one for the image of bands at 20-m spacing, and another one for the image of bands at 10-m spacing. We will sample patches of size 8×8 for the image at 20-m spacing, and patches of size 16×16 for the image at 10-m spacing. We set the **OTB_TF_NSOURCES** environment variable to *2* in order to tell the OTBTF applications that we want 2 sources (one for the 10-m spacing image, one for the 20-m spacing image).

Run the following command:

```
# Here, we change the number of sources
export OTB_TF_NSOURCES=2

# Patches extraction
otbcli_PatchesExtraction \
-source1.il Sentinel-2_B567_20m.tif \
-source1.patchsizex 8 \
-source1.patchsizey 8 \
-source1.out Sentinel-2_B567_20m_patches_A.tif \
-source2.il Sentinel-2_B4328_10m.tif \
-source2.patchsizex 16 \
-source2.patchsizey 16 \
-source2.out Sentinel-2_B4328_10m_patches_A.tif \
-vec terrain_truth_epsg32654_A_pos.shp \
```

```
-field "class" \
-outlabels Sentinel-2_B567_20m_labels_A.tif uint8
```

Now, do the same for the validation dataset:

```
# Patches extraction
otbcli_PatchesExtraction \
-source1.il Sentinel-2_B567_20m.tif \
-source1.patchsizex 8 \
-source1.patchsizey 8 \
-source1.out Sentinel-2_B567_20m_patches_B.tif \
-source2.il Sentinel-2_B4328_10m.tif \
-source2.patchsizex 16 \
-source2.patchsizey 16 \
-source2.out Sentinel-2_B4328_10m_patches_B.tif \
-vec terrain_truth_epsg32654_B_pos.shp \
-field "class" \
-outlabels Sentinel-2_B567_20m_labels_B.tif uint8
```

8.5 Training

Our patches are ready. We use the **TensorflowModelTrain** for training our new network. Use the two placeholders for the input image patches, both for training and validation sources.

Run the following command:

```
# Train the deep learning model
otbcli_TensorflowModelTrain \
-training.source1.il Sentinel-2_B567_20m_patches_A.tif \
-training.source1.patchsizex 8 \
-training.source1.patchsizey 8 \
-training.source1.placeholder "x1" \
-training.source2.il Sentinel-2_B4328_10m_patches_A.tif \
-training.source2.patchsizex 16 \
-training.source2.patchsizey 16 \
-training.source2.placeholder "x2" \
-training.source3.il Sentinel-2_B567_20m_labels_A.tif \
-training.source3.patchsizex 1 \
-training.source3.patchsizey 1 \
-training.source3.placeholder "y" \
-model.dir model3 \
-training.targetnodes "optimizer" \
-validation.mode class \
-validation.source1.il Sentinel-2_B567_20m_patches_B.tif \
-validation.source1.name "x1" \
-validation.source2.il Sentinel-2_B4328_10m_patches_B.tif \
-validation.source2.name "x2" \
-validation.source3.il Sentinel-2_B567_20m_labels_B.tif \
-validation.source3.name "prediction" \
-model.saveto model3/variables/variables
```

8.5.1 Inference

In this section, we will generate some maps.

8.5.1.1 Patch-based mode

First, we will run the model in patch-based mode. Generate the output classification map at the resolution of the 20-m-spacing image. To do this, just feed the 20-m-spacing image to the **source1**, which is the reference source for the output physical spacing.

```
# Generate the classification map, patch-based mode, same
# resolution as the 20m spacing image
otbcli_TensorflowModelServe \
-source1.il Sentinel-2_B567_20m.tif \
-source1.rfieldx 8 \
-source1.rfieldy 8 \
-source1.placeholder "x1" \
-source2.il Sentinel-2_B4328_10m.tif \
-source2.rfieldx 16 \
-source2.rfieldy 16 \
-source2.placeholder "x2" \
-model.dir model3 \
-output.names "prediction" \
-out "classif_model3.tif?&box=2000:2000:500:500"
```

If you switch **source1** and **source2** so the first source is the 10-m-spacing image, then the output image will be created with the same spacing. Let's try it!

8.5.1.2 Fully convolutional mode

Generate the output classification map at the same resolution as the 10-m-spacing image, setting the **fullyconv** parameter to **on** to enable the fully convolutional mode of the processing.

Use the following command line:

```
# Generate the classification map, fully convolutional
# mode, same resolution as the 20m spacing image
otbcli_TensorflowModelServe \
-source1.il Sentinel-2_B567_20m.tif \
-source1.rfieldx 8 \
-source1.rfieldy 8 \
-source1.placeholder "x1" \
-source2.il Sentinel-2_B4328_10m.tif \
-source2.rfieldx 16 \
-source2.rfieldy 16 \
-source2.placeholder "x2" \
-model.dir model3 \
-model.fullyconv on \
-output.names prediction \
-out "classif_model3_fcn.tif?&box=2000:2000:500:500"
```

> **Important note**
> If you switch **source1** and **source2** so the first source is the 10-m-spacing image, you have to provide an output scale factor (parameter **spcscale**) to the **TensorflowModelServe** application, to make consistent the physical spacing of the output. In this example, the **spcscale** would be 4.

9

Discussion

In this section, we have introduced a simple approach to produce land cover maps from remote sensing images using deep nets. The patch-based approach, consisting of estimating the class value of one patch of image(s), is well suited to a sparsely annotated dataset. Architectures involved in patch-based classification generally input one or multiple patches of images, possibly from different modalities (as we saw in section 8.2), and derive a single-valued output corresponding to the estimated class of the input content. This kind of architecture can be trained with single-valued class samples. For instance, terrain truth data, like a set of GPS coordinates associated with land cover classes, suffices to train this kind of network. In chapter 7 we saw that these models are also good feature extractors, that can be used in other classifiers like Random Forests. We explain why, depending on their implementation, models can be faster at inference time, and introduced fully convolutional patch-based models in section 6.2.

We also show that the patch-based approach has some drawbacks. In particular, the receptive field of one model has an impact on the spatial resolution of the output map generated with the trained model. Since the image content in the receptive field activates the single-valued model output, we can expect that the localization of the predicted class is not reliable if predicted at the same spatial resolution as the input image. Practically, large patches contain more context, so the network can be activated by a class target (i.e. a tree, a road), even if this last is not exactly centered on the patch. We can explain theoretically this phenomenon analyzing the constitution of the patch-based models. For instance, VGG-like architectures [24] are composed of a succession of pooling operations and convolutions with strides. These operations downsample the feature maps in the spatial dimension, through the network depth, generally with a factor of 2. Each downsampling diminishes the spatial resolution of the features, causing a loss of localization of the predicted classes, and usually the last feature maps of this kind of network have a low spatial resolution, corresponding to 2^n times the physical spacing of the input. In practice, when a land cover map is generated at the same resolution as the input image, rather than the resolution of the last feature map of the network (which is, generally 2^n smaller), some blobby artifacts can be observed around small objects. In the next section, we will introduce some new architectures that aim to preserve the spatial resolution of the final features of the model output.

Part III

Semantic segmentation

10

Semantic segmentation of optical imagery

10.1 Introduction

This section presents an approach to perform semantic segmentation on very-high-resolution remote sensing images using a deep convolutional neural network. In part II, we introduced a patch-based classification approach, that is suited to sparsely annotated data. However, this kind of approach is limited in term of semantic precision in spatial location. Indeed, a patch-based network is designed to attribute one single class to an input patch, and does not make use of the context of the actual classes labels, namely the terrain truth. On the contrary, semantic segmentation methods, also known as *dense prediction*, use the semantic spatial context. These approaches train networks estimating the semantic of patches of pixels, rather than just a single pixel. Semantic segmentation methods allow us to tackle the spatial precision of the classification. However, the training of networks requires densely annotated terrain truth. In this section, we will introduce a semantic segmentation approach, and use OpenStreetMap data over the city of Amsterdam to generate the map of buildings from a Spot image (1.5-m resolution).

10.2 Overview

The goal of semantic image segmentation is to label each pixel of an image with a corresponding class. The main difference with pixel classification, as performed in the previous sections, is that the classes are estimated for every pixel in an entire region of an image, rather than one pixel by one pixel (fig. 10.1). That is why semantic segmentation is also referred as dense prediction.

FIGURE 10.1
Semantic segmentation is understanding an image at pixel level. The output is an image of labels that annotate the image content: here the output of the semantic segmentation is a label image annotating the following 3 classes: human, bike and background. *Source: VOC2012.*

As for image classification, CNNs are very successful on segmentation problems. First approaches using artificial neural networks used fully connected layers and therefore required fixed size images. Later, FCN CNN architectures for dense predictions were usable without any fully connected layers. One of the main problems remaining with using CNNs for semantic segmentation is pooling layers. Pooling layers increase the field of view and are able to aggregate the context, but as we saw in section 6, they remove spatial information. Since semantic segmentation requires a precise location of classes, it needs to preserve this spatial information. To tackle this problem, many architectures have evolved over the last years. In this chapter, we will use one of the first popular architectures based on FCN, the U-Net [25]. It is an encoder-decoder architecture. The encoder progressively reduces the spatial resolution with strided convolution or pooling layers, increasing the feature dimension, and the decoder progressively recovers spatial resolution in decreasing feature dimension. Finally, the last feature map is the predicted class. Figure 10.2 presents the original U-Net architecture.

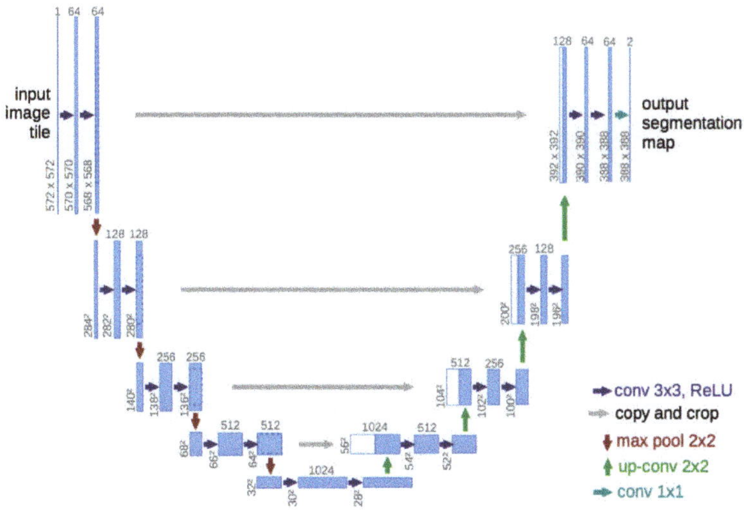

FIGURE 10.2
The original U-Net architecture [25]. The first half of the network, referred to as the encoder, downsamples the input image and feature maps. The second half of the network, called the decoder, aims to recover features at higher resolutions, in upsampling the input feature maps and concatenating them with the encoder feature maps of the same resolution (also called *skip connections*).

11

Data used: the Amsterdam dataset

This section aims to present the data involved for the upcoming practice session. We rely on freely available data, called the *Amsterdam* dataset. In the following part, we explain how to download and prepare this dataset in order to be ready for the practice session. Create a directory named *amsterdam* on your file system. We will use this directory to store all the downloaded and generated data during the exercises.

11.1 Description

The *Amsterdam* dataset consists of one very-high-resolution Spot-7 image acquired over the city of Amsterdam, and a vector layer containing the building footprints. In this section, we explain how to download the Spot-7 image, and generate the vector layer from OSM data. The *Amsterdam* dataset is involved in the entire section 11. Note that you can generalize this exercise to any orthorectified very-high-resolution remote sensing image at your disposal.

11.2 Spot-7 image

The Spot-7 product is composed of one multispectral image (6.0-m physical spacing, bands: red, green, blue, near infra-red) and one panchromatic channel (1.5-m physical spacing). The encoding is 12 bits. The Spot image can be freely downloaded from the Airbus, Defense & Space sample imagery web page.[1] To download the product, follow these steps:

1. Go to the ADS sample imagery web page.
2. In the search field, enter *amsterdam* (next to the magnifying icon).
3. Click on the *Search* button.

[1] `https://www.intelligence-airbusds.com/en/8262-sample-imagery`

TABLE 11.1
Characteristics of the Spot-7 product

Id	Mode	*col* × *row* × *band*	Encoding	Pixel spacing
SPOT7_P	Pan	6616 × 5824 × 1	12 bits	1.5 m
SPOT7_MS	MS	1654 × 1456 × 4	12 bits	6.0 m

4. Select the available Spot-7 product named **SPOT 7 - Amsterdam, Netherlands.**

Then, download the Spot-7 image from the link provided. Once the download is finished, unzip the archive in your *amsterdam* directory, in a sub-directory named *Amsterdam_Netherlands_SPOT7*. Then, take a look at the contents of the Spot product directory. You can notice that there are two separate directories in **PROD_SPOT7_001/VOL_SPOT7_001_A** :

- Sub-directory **IMG_SPOT7_MS_001_A**: Contains the multispectral imagery in raster **IMG_SPOT7_MS_201409171025192_ORT_1190912101_R1C1.JP2**.
- Sub-directory **IMG_SPOT7_P_001_A**: Contains the panchromatic channel in raster **IMG_SPOT7_P_201409171025192_ORT_1190912101_R1C1.JP2**.

Image characteristics are reported in table 11.1. Both images are stored in the JPEG2000 format, with the coordinate reference system EPSG 32631 (Zone 31 North).

Import the Spot-7 image in QGIS from the menu: **Layer** > **Add layer** > **Add raster layer**, then select either the panchromatic channel or the multispectral image of the downloaded Spot-7 product. In the next section, we will use the satellite image only to select the region of interest where we need to retrieve the OSM data.

11.3 OpenStreetMap data

The following steps explain how to download the OSM data, and prepare a vector layer containing polygons representing building footprints. All operations are performed using QGIS.

Skip the OSM data preparation
Note that you can skip the steps for the OSM vector data generation (including the rasterization step described in section 12.1.4), and download directly the label image of the OSM building footprints prepared

by the author, from the resources repository[a] (use the **Find file tool** to search for the *amsterdam_labelimage.tif* file).

[a]`https://github.com/remicres/otbtf_tutorials_resources`

11.3.1 OSM downloader plugin

The OSM plugin is a user-friendly tool that enables you to download OpenStreetMap data over a region of interest.

1. In the menu, select **Plugins** > **Manage and install plugin...**
2. Search for *OSM Downloader* as shown in figure 11.1.
3. Click on **Install plugin.**

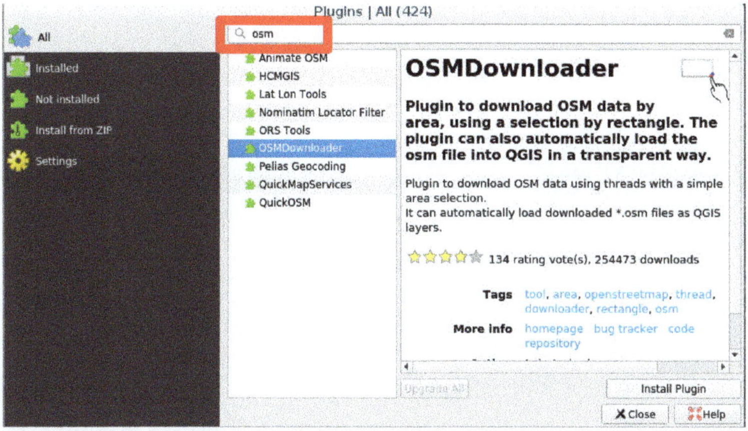

FIGURE 11.1
Install OSM Downloader.

Once the plugin is installed, one can retrieve the OSM data.

11.3.2 Download OSM data

1. Click on **Download OSM data by rectangle selection** (figure 11.2).
2. Select an area covering the central part of the image (figure 11.3). Do not select an area too large, else the download might fail. A region corresponding approximately to half the image is enough.
3. Select an output filename on your *amsterdam* directory as *amsterdam.osm.*
4. Tick the box **Load layer after download.**

5. Click on **Ok**.

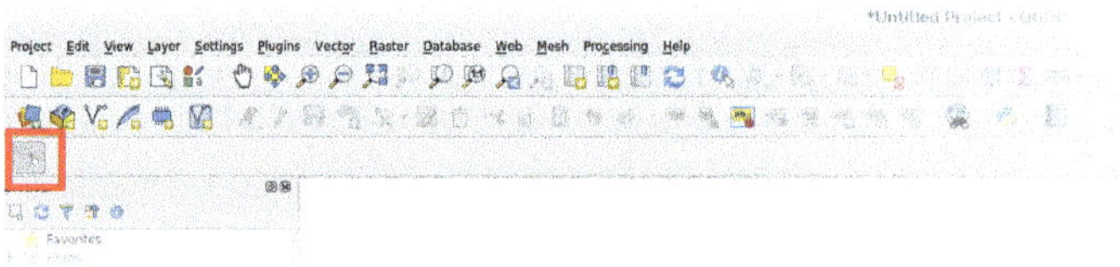

FIGURE 11.2
Use the OSM downloader from the QGIS toolbar.

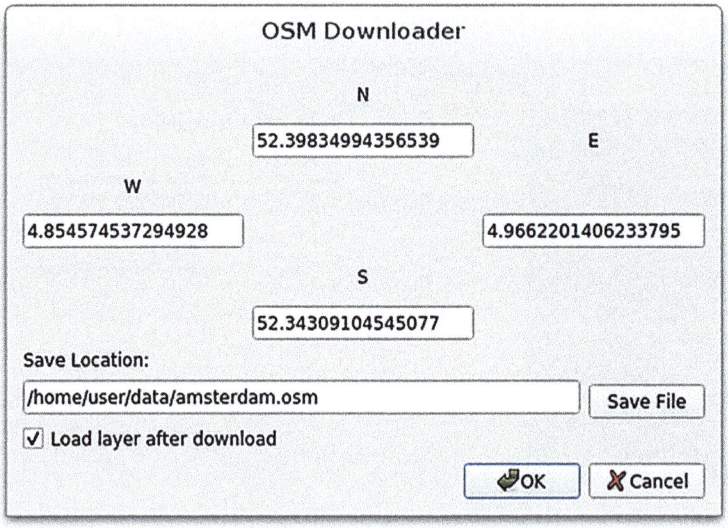

FIGURE 11.3
Download OSM data by rectangle selection.

After the download, the vector layer must be imported in QGIS. Select the **Multipolygon** layer ref to fig 11.4.

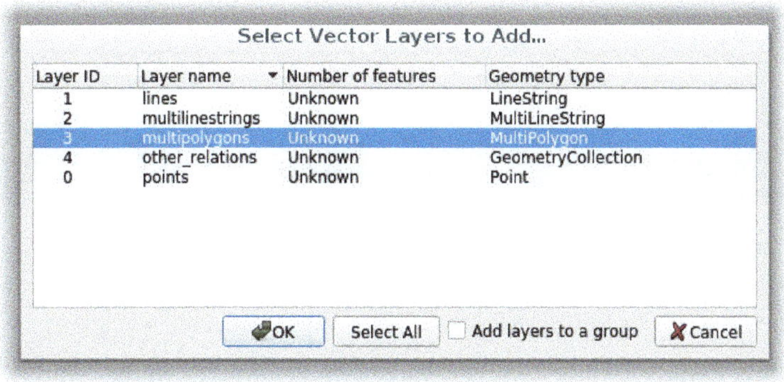

FIGURE 11.4
Import only the multipolygon layer.

11.3.3 Prepare the vector layer

We will now export a vector layer for the buildings in the Shapefile format from the OSM file.

1. Right-click on the imported **multipolygon_layer** and select **Open attribute table** (figure 11.5).
2. Click on the **Filter by attribute** icon (figure 11.6).
3. Enter the following expression: ***length("buildings")>0***, to select only buildings polygons (figure 11.7), then click on **Select features**. Then click on the **Close** button.
4. Right-click on the **multipolygon_layer** and select **Export** > **Save selected features as** (figure 11.8).
5. Choose a name for your output vector layer: **buildings.gpkg**. Select also the same coordinate reference system as the Spot-7 image, **EPSG 32631** (figure 11.9).

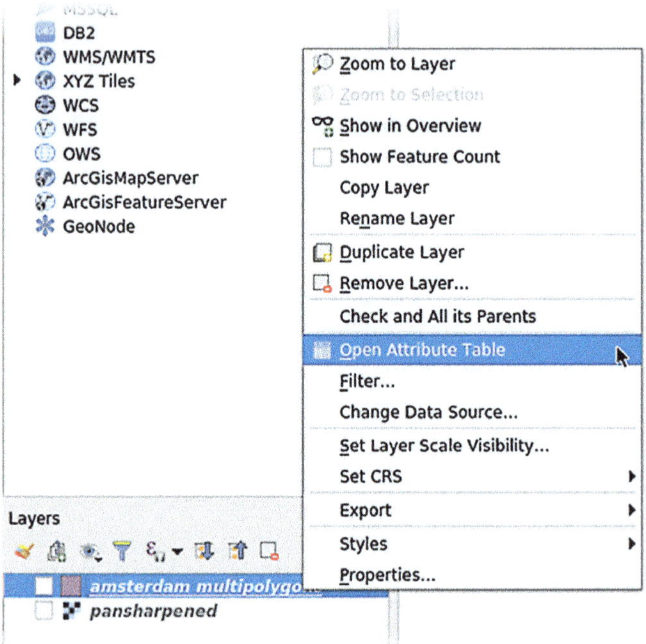

FIGURE 11.5
Open the attribute table.

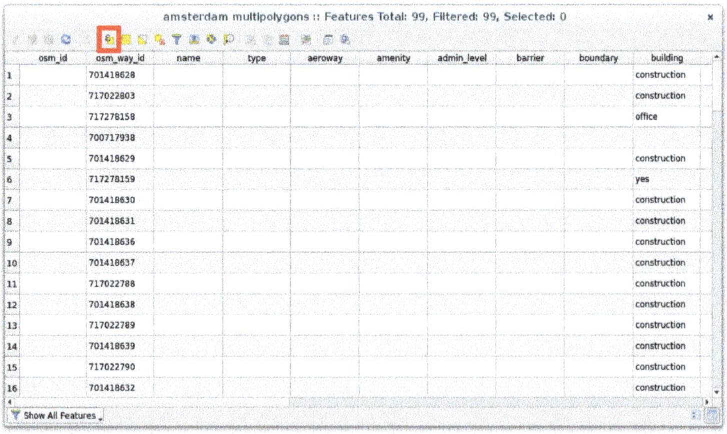

FIGURE 11.6
Click on the **Filter by attribute** icon.

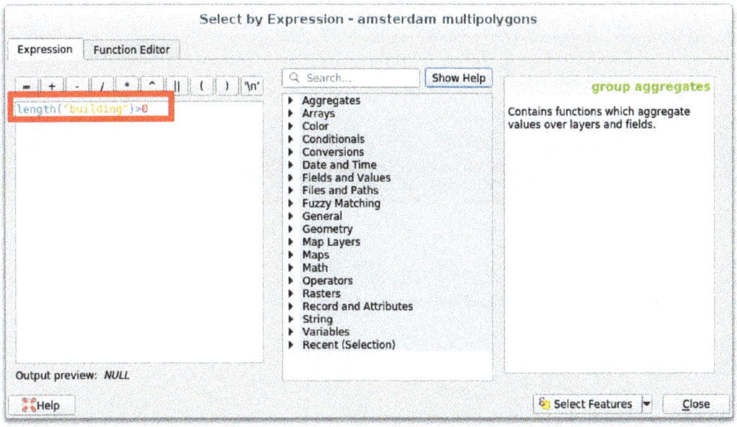

FIGURE 11.7
Select buildings polygons.

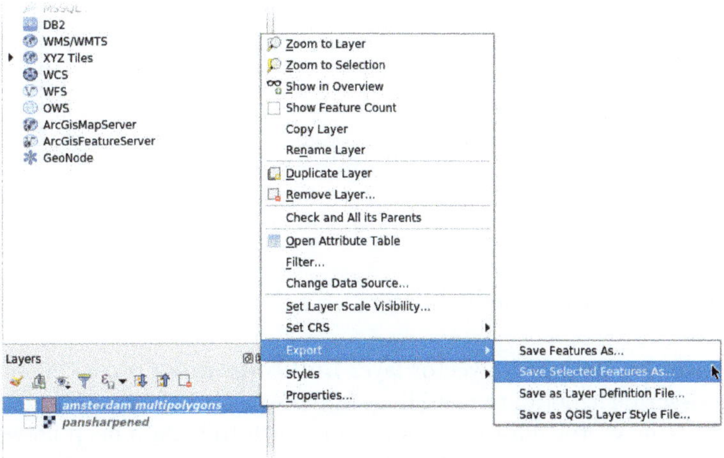

FIGURE 11.8
Export the selected polygons as a new vector layer.

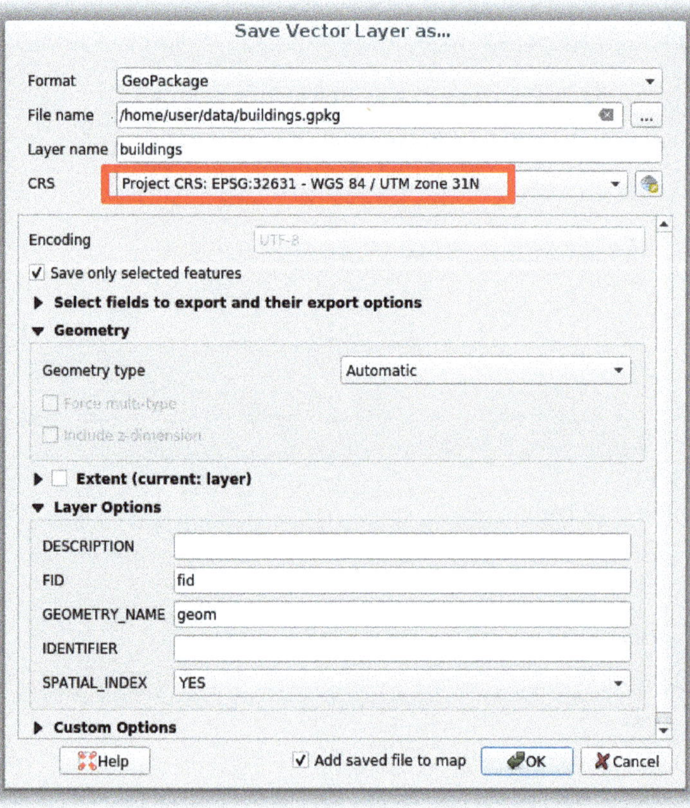

FIGURE 11.9
Export the output vector layer.

You have now prepared a vector layer **buildings.gpkg** consisting of polygons that describe in detail the buildings of the Amsterdam area (figure 11.10). We will use this vector data later as terrain truth to train a deep network to perform the semantic segmentation of the buildings.

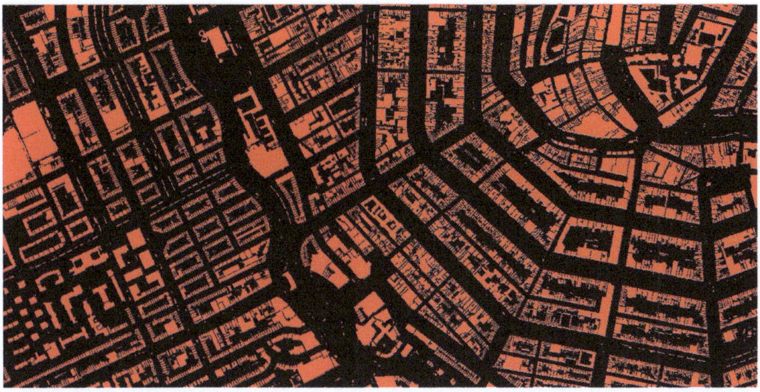

FIGURE 11.10
The resulting **buildings.gpkg** vector layer (in red).

OSM downloads
Instead of using the *OSM Downloader* plugin from QGIS, which might be limited by the amount of data to download, you can use several alternatives, such as the following:

- dowload.geoafabrik.de
- OpenStreetMap.org/export

12

Mapping buildings

In this section, we apply semantic segmentation to map the buildings over the Amsterdam region of interest.

12.1 Input data pre-processing

We intend to perform the semantic segmentation of buildings, from a Spot-7 image at 1.5 m acquired over Amsterdam. The first step is to pre-process the Spot-7 image. Then, we will prepare the terrain truth that will be used for training of our U-Net model. The main difference with the approach investigated in Part II, is that the labels are patches rather than a single label value. Hence we need to produce a label image for our terrain truth, with the same pixel size of the Spot-7 image.

12.1.1 Satellite image pansharpening

For the sake of simplicity, we will introduce a very simple deep network that inputs a single image. For this reason, we will fuse the panchromatic channel and the multispectral image into one single image, prior to other processing. We perform pansharpening of the Spot-7 product, meaning that we generate a multispectral image that has the same resolution as the panchromatic channel, and the same channels as the original multispectral image. To generate this pansharpened image, we rely on the **BundleToPerfectSensor** application of OTB.

```
# Perform the pansharpening

# Go in the folder of the Spot product
cd ~/amsterdam/
cd Amsterdam_Netherlands_SPOT7/
cd PROD_SPOT7_001/VOL_SPOT7_001_A/

# Execute the OTB application
# We can set the output encoding to uint8 since inputs encoding is
# 8 bits.
otbcli_BundleToPerfectSensor \
-inp IMG_SPOT7_P_001_A/IMG_SPOT7_P_201409171025192_ORT_1190912101_R1C1
 .JP2 \
```

```
-inxs IMG_SPOT7_MS_001_A/IMG_SPOT7_MS_201409171025192_ORT_1190912101
 _R1C1.JP2 \
-method bayes \
-out ~/amsterdam/IMG_SPOT7_amsterdam_PMS.tif uint8

# Go back in the workspace directory
cd ~/amsterdam/
```

12.1.2 Image normalization

We normalize the pansharpened Spot-7 image between 0 and 1 using the **BandMathX** application.

```
# Normalize image between [0,1]
otbcli_BandMathX \
-il "IMG_SPOT7_amsterdam_PMS.tif" \
-exp "{im1b1/im1b1Maxi;im1b2/im1b2Maxi;␣\
im1b3/im1b3Maxi;im1b4/im1b4Maxi}" \
-out "IMG_SPOT7_amsterdam_PMS_norm.tif"
```

12.1.3 Sample selection

In the following, we will select the center of the patches that we will extract in the images, and that will be used as the terrain truth.

12.1.3.1 Patch position seeding

The goal here, is to select only patches that are fully annotated: in our case, if a patch has missing building annotations, we remove it from the terrain truth. Typically, in the case of buildings, we should avoid terrain truth patches that miss buildings that actually exist in Spot-7 patches. In order to select a sufficient number of patches for the terrain truth, we adopt a simple strategy consisting of selecting patch positions at regular intervals in the images. Since we want fully annotated patches, we just keep the positions that are fully included in the area covered by the OSM data, in our case the region of interest used in section 11.3. Our goal is to generate a vector layer containing the locations of the patches that will be sampled.

Here are the steps to select them:

1. Use QGIS to create a vector grid. Select from the menu **Vector** > **Research tools** > **Create grid** as shown in figure 12.1.
2. Create a grid with the following properties (figure 12.2):
    - Grid type: *Rectangle*.
    - Horizontal spacing and Vertical spacing: 96 meters.
    - Grid extent: select *Use extent from* then use the Spot-7 image layer to specify the extent.
    - Name the output vector data *amsterdam_grid.gpkg*.

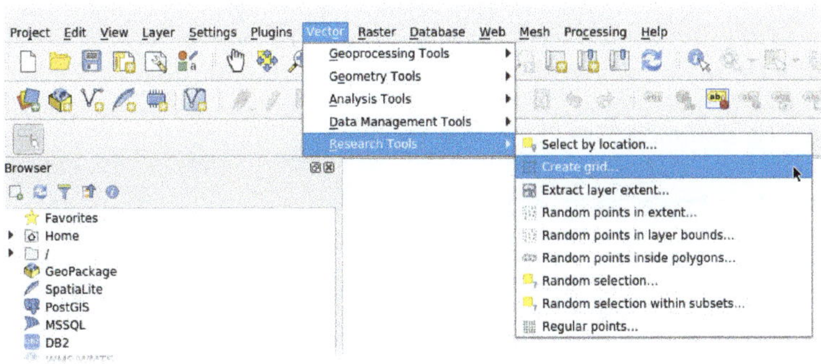

FIGURE 12.1
Create a vector grid.

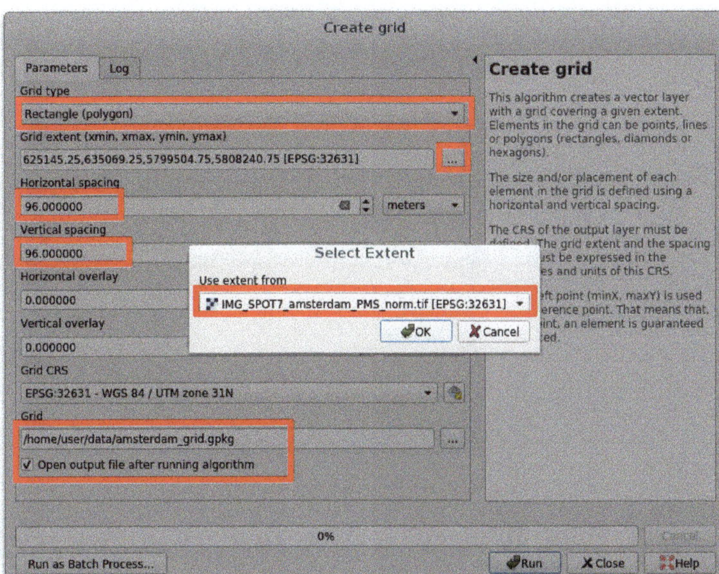

FIGURE 12.2
Set the grid properties.

3. Once the grid is created, select it in the stack list on the side of the QGIS window.
4. Click on the *Select features by area or single click* icon of the QGIS toolbar, as shown in figure 12.3.
5. Select all the polygons of the grid which lie completely inside the area covered by the buildings polygons.

6. Finally, you can export the selected polygons: right-click on the grid layer and select **Export > Save selected features as**, then export the vector data as *amsterdam_grid_selection.gpkg*.

Figure 12.4 summarizes these last two steps.

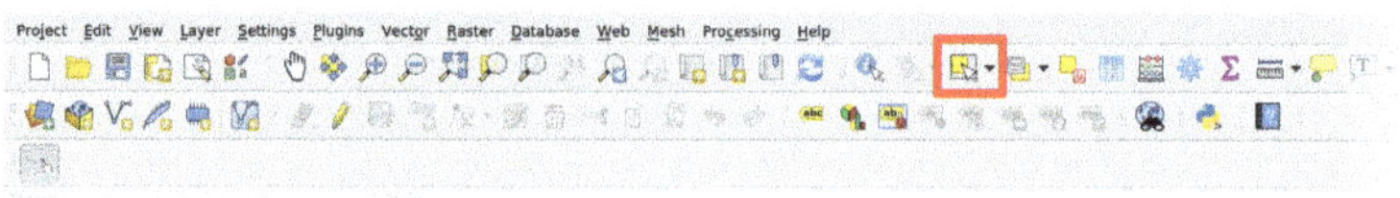

FIGURE 12.3
The *Select features by area or single click* icon of the QGIS toolbar.

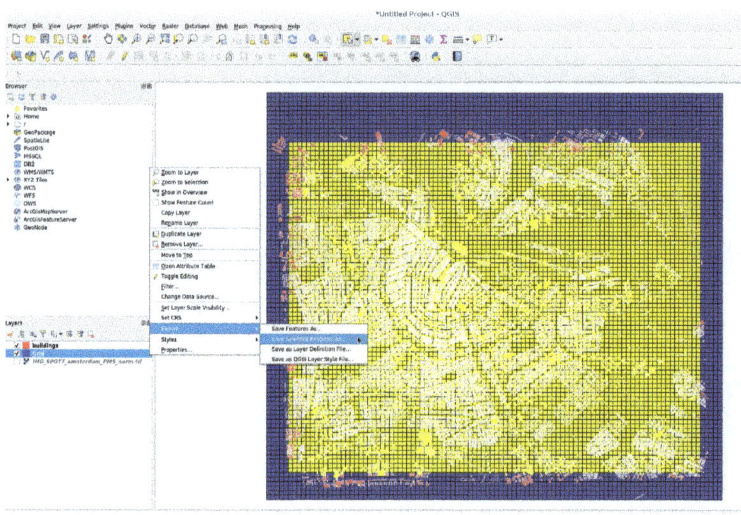

FIGURE 12.4
Select the patches that lie completely inside the area covered with building annotations from OSM (the selection appears in yellow in the illustration), and export the selection as a new vector layer, *amsterdam_grid_selection.gpkg*.

Sometimes, the OSM vector data is not fully annotated and many polygons are missing in some regions. In this case, you can manually clean the vector data, removing positions of patches that are not actually fully annotated. This is a common issue that you can encounter with OSM data. Luckily, OSM data is quite well annotated in our *Amsterdam* dataset.

12.1.3.2 Patch position selection

Now, we want to extract centroids of those valid patches that we previously have pre-selected. From the QGIS menu, select **Vector**, then **Geometry tools**, then **Centroids** (Fig. 12.5).

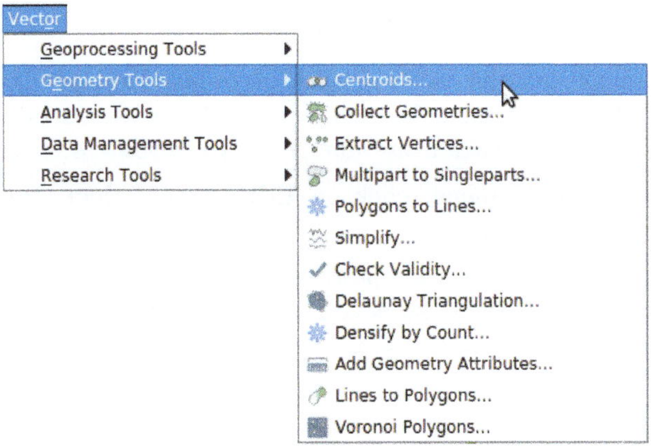

FIGURE 12.5
Select the centroids computation in QGIS.

Select *amsterdam_grid_selection.gpkg* as the input layer and save the output vector layer for the centroids in *amsterdam_centroids.gpkg*.

12.1.3.3 Patches split

Now we select two distinct groups of patches, for the same reason as explained in section 5.2: one for the training and another for the validation. Select 70% of the patches for the training dataset using QGIS: from the menu, select **Vector** > **Research tools** > **Random Selection** (Fig. 12.6). Select *amsterdam_centroids.gpkg* as the **input layer**, and select the **method** as **percentage of selected features**. Create a selection with 70% of the patch centroids.

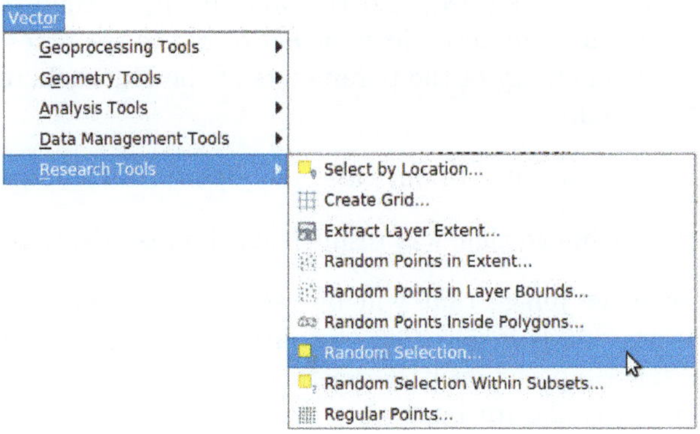

FIGURE 12.6
Select some centroids randomly using QGIS.

Then, perform the following steps in QGIS:

1. Right-click on the *sane_patches_centroids.shp* layer (left panel called **Layers**) and select **Export** > **Save selected features as...** (Fig. 12.7). Save it as *sane_patches_centroids_A.shp.*
2. From the QGIS menu, click on **Edit**, then **Select**, then **Invert feature selection.**
3. Right-click on the *sane_patches_centroids.shp* layer and select **Export**, then **Save selected features as...**, and save it as *sane_patches_centroids_B.shp.*

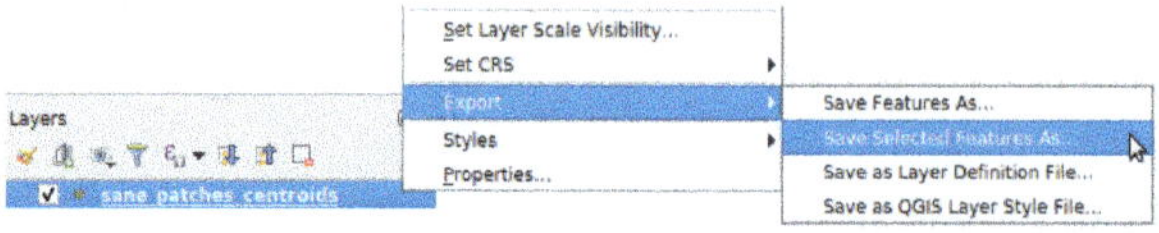

FIGURE 12.7
Select some centroids randomly using QGIS.

Now we should have two vector layers that describe the location of the center of the patches to sample, which are mutually exclusive (i.e. no point is shared in two groups).

12.1.4 Rasterization

We will use the available vector data from OpenStreetMap, which we will rasterize at the same spatial extent and resolution of the Spot-7 image. For this, we use the OTB application named **Rasterization**. We remind the reader that this section can be skipped, and the label image of the OSM buildings that we prepared can be directly downloaded from the resources repository.[1]

Here is a short summary of the parameters of the **Rasterization** application that will be used:

- **in** is the input vector data. In our case, the OpenStreetMap polygons.
- **out** is the output image that will result from the rasterization process.
- **im** is the reference image that will be used as the reference grid for the rasterization process. It means that the output image (parameter **out**) will be perfectly superimposed over the reference image. In our case, we will set the Spot-7 image in the **im** image.
- **mode** is the processing mode: we chose **attribute** mode to burn the values contained in the *class* field of the vector data.

[1] `https://github.com/remicres/otbtf_tutorials_resources`

   - **mode.attribute.field** is the name of the field from which we want to burn the values. In our case, we set *class.*

- **background** is the value for the background. In our case, we set the default value (zero) because there is no zero value in the *class* field.

Let's run the application.

```
# Rasterize the OSM vector layer for terrain truth
# Since the output raster contains integers values
# between 0 and 6, we can use uint8 encoding (8 bits)
otbcli_Rasterization \
-in "buildings.gpkg" \
-im "IMG_SPOT7_amsterdam_PMS_norm.tif" \
-out "amsterdam_labelimage.tif" uint8 \
-mode binary \
-mode.binary.foreground 1
```

You can now open the *amsterdam_labelimage.tif* in QGIS to check that the vector layer has been correctly rasterized over the Spot-7 image pixel grid, as shown in figure 12.8.

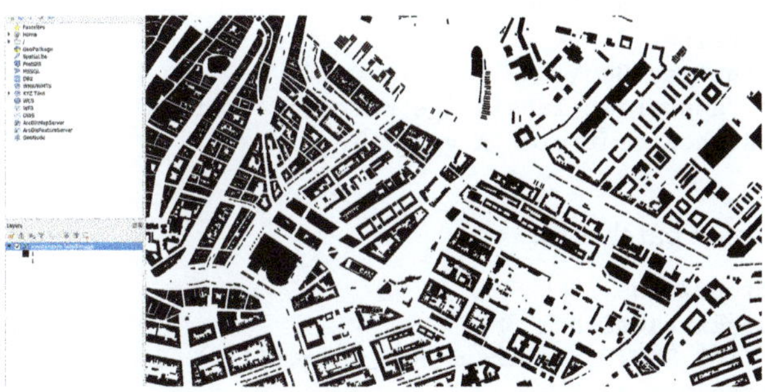

FIGURE 12.8
Rasterized building polygons over Spot-7 image reference.

12.1.5 Patch extraction

Let's prepare the patches that will be used for training and validation. We use the **PatchesExtraction** application to jointly extract patches in the Spot-7 image and the label image. In Part II, each remote sensing image patch was associated with a corresponding label value. This time, our goal is slightly different because we need labels as patches, not as a single value like for the patch-based approach. Hence, we tell **PatchesExtraction** that we want two sources: one for the Spot-7 image, and one for the label image. We change the **OTB_TF_NSOURCES** environment variable to 2 to fulfill this need.

```
export OTB_TF_NSOURCES=2
```

We extract the patches in positions that we selected in section 12.1.3, at the positions specified respectively in:

1. *sane_patches_centroids_A.shp* layer for training patches,
2. *sane_patches_centroids_B.shp* layer for validation patches.

We choose patches of size 64×64 pixels since it should be enough to train our small network.

In the command line, this should look like:

```
# Here, we change the number of sources to 2, because
# we need one source for the Spot-7 image and one
# source for the label image
export OTB_TF_NSOURCES=2

# Patches extraction
otbcli_PatchesExtraction \
-source1.il "IMG_SPOT7_amsterdam_PMS_norm.tif" \
-source1.patchsizex 64 \
-source1.patchsizey 64 \
-source1.out "amsterdam_Spot7_patches_A.tif" \
-source2.il "amsterdam_labelimage.tif" \
-source2.patchsizex 64 \
-source2.patchsizey 64 \
-source2.out "amsterdam_labelimage_patches_A.tif" \
-vec "sane_patches_centroids_A.gpkg" \
-field "id"
```

Do it then for patches B:

```
# Patches extraction
otbcli_PatchesExtraction \
-source1.il "IMG_SPOT7_amsterdam_PMS_norm.tif" \
-source1.patchsizex 64 \
-source1.patchsizey 64 \
-source1.out "amsterdam_Spot7_patches_B.tif" \
-source2.il "amsterdam_labelimage.tif" \
-source2.patchsizex 64 \
-source2.patchsizey 64 \
-source2.out "amsterdam_labelimage_patches_B.tif" \
-vec "sane_patches_centroids_B.gpkg" \
-field "id"
```

You can then open your patch images in QGIS and check them visually. Our training and validation data are now ready!

12.2 Building the model

We have prepared a small U-Net model with 4 downscaling/upscaling levels.

12.2.1 Architecture

This kind of architecture is also called an autoencoder. An autoencoder consists of:

1. an encoder extracting features in the input image,
2. a bottleneck, being the output of the encoder, and
3. a decoder that upsamples the features into the output features (typically for a classifier: the classes representation).

Figure 12.9 presents our model.

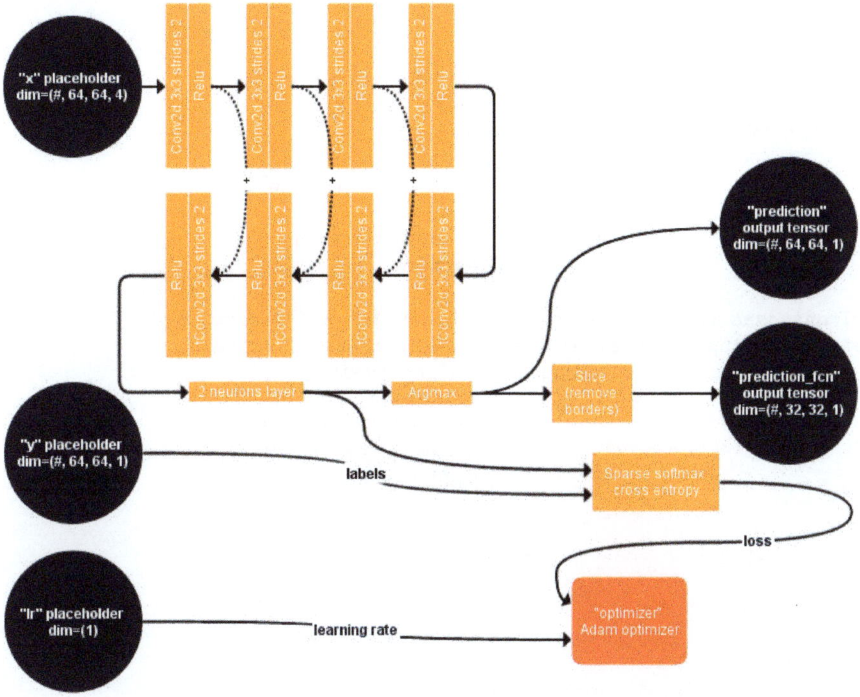

FIGURE 12.9
Our U-net model with 4 upsampling/downsampling levels.

The encoder is composed of 4 convolutions with stride 2 followed with a rectified linear unit. The decoder is composed of 4 transposed convolutions with stride 2 with a rectified linear unit. The final layer is a 2-neuron layer: one neuron for class *background* and another for class *buildings*. We train the network in minimizing the gap between the predicted (i.e. the 2 neurons output) and the terrain truth (i.e. the y placeholder which hosts the label image patches).

12.2.2 Implementation

Let's look a bit more closely at the code:

```
#
# Do not forget to add the otbtf/python directory to your PYTHONPATH.
# Else "tricks" can't be imported.
# In the OTBTF Docker image, tricks.py is located in the following
# path: /work/otb/otb/Modules/otbtf
#
# Works with OTBTF 1.7
#
from tricks import *
import sys

nclasses=2

# Convolution block with strides 2 ("downsampling")
def _conv(inp, n, k_size=3, strides=2, activ=tf.nn.relu):
  return tf.layers.conv2d(inputs=inp,
                          filters=n,
                          kernel_size=[k_size, k_size],
                          padding="same",
                          strides=(strides, strides),
                          activation=activ)

# Transposed convolution block with strides 2 ("upsampling")
def _dconv(inp, n, k_size=3, strides=2, activ=tf.nn.relu):
  return tf.layers.conv2d_transpose(inputs=inp,
                                    filters=n,
                                    strides=(strides,strides),
                                    kernel_size=[k_size, k_size],
                                    padding="same",
                                    activation=activ)

def myModel(x):

  depth = 16

  # Encoding
  conv1 = _conv(x, 1*depth) # 64 x 64 --> 32 x 32 (31 x 31)
  conv2 = _conv(conv1, 2*depth) # 32 x 32 --> 16 x 16 (15 x 15)
  conv3 = _conv(conv2, 4*depth) # 16 x 16 --> 8 x 8 ( 7 x 7)
  conv4 = _conv(conv3, 4*depth) # 8 x 8 --> 4 x 4 ( 3 x 3)

  # Decoding (with skip connections)
  deconv1 = _dconv(conv4, 4*depth) # 4 x 4 --> 8 x 8 ( 5 x 5)
  deconv2 = _dconv(deconv1 + conv3, 2*depth) # 8 x 8 --> 16 x 16
                                             # ( 9 x 9)
  deconv3 = _dconv(deconv2 + conv2, 1*depth) # 16 x 16 --> 32 x 32
                                             # (17 x 17)
  deconv4 = _dconv(deconv3 + conv1, 1*depth) # 32 x 32 --> 64 x 64
                                             # (33 x 33)

  # Neurons for classes
```

```
  estimated = tf.layers.dense(inputs=deconv4, units=nclasses,
                              activation=None)

  return estimated

""" Main """
# check number of arguments
if len(sys.argv) != 2:
  print("Usage␣:␣<output␣directory␣for␣SavedModel>")
  sys.exit(1)

# Create the graph
with tf.Graph().as_default():

  # Placeholders
  x = tf.placeholder(tf.float32, [None, None, None, 4], name="x")
  y = tf.placeholder(tf.int32 , [None, None, None, 1], name="y")
  default_lr = tf.constant(0.0002, dtype=tf.float32, shape=[])
  lr = tf.placeholder_with_default(default_lr, shape=[], name="lr")

  # Output neurons of the model
  y_estimated = myModel(x)

  # Prediction output
  y_class = tf.argmax(y_estimated, axis=3)

  # Add the 4th dimension to have (#, :, :, 1) so the OTBTF apps
  # know that the output image has 1 component
  y_pred = tf.expand_dims(y_class, axis=-1, name="prediction")

  # Prediction output (FCN)
  y_pred_fcn = tf.identity(y_pred[:, 16:-16, 16:-16, :],
                           name="prediction_fcn")

  # Loss function
  cost = tf.losses.sparse_softmax_cross_entropy(labels=tf.reshape
         (y, [-1, 1]), logits=tf.reshape(y_estimated, [-1, nclasses]))

  # Optimizer
  optimizer = tf.train.AdamOptimizer(learning_rate=lr,
              name="optimizer").minimize(cost)

  # Initializer, saver, session
  init = tf.global_variables_initializer()
  saver = tf.train.Saver( max_to_keep=20 )
  sess = tf.Session()
  sess.run(init)

  # Create a SavedModel
  CreateSavedModel(sess=sess,
                   inputs=["x:0","y:0"],
                   outputs=["prediction:0", "prediction_fcn:0"],
                   directory=sys.argv[1])
```

As can be noticed, we use convolutions and transposed convolutions which retain the size of the output features (*padding="same"*), meaning that zero padding is used in convolutions. We should understand that we can't use the features computed in the last layer directly, if we want to use this model in fully convolutional mode: for each processed block, only a subset would be *exact*, meaning similar, regardless of the input region. Using the output would produce some blocking artifacts caused by the zero padding in the convolutions.

12.2.2.1 Exact output

To cope with that, we create an output dedicated to the computation of the *exact* part of the *prediction* output tensor. For this, we simply remove a few pixels from the borders of the output since the convolutions "pollute" the borders at each layer, as illustrated in figure 12.10.

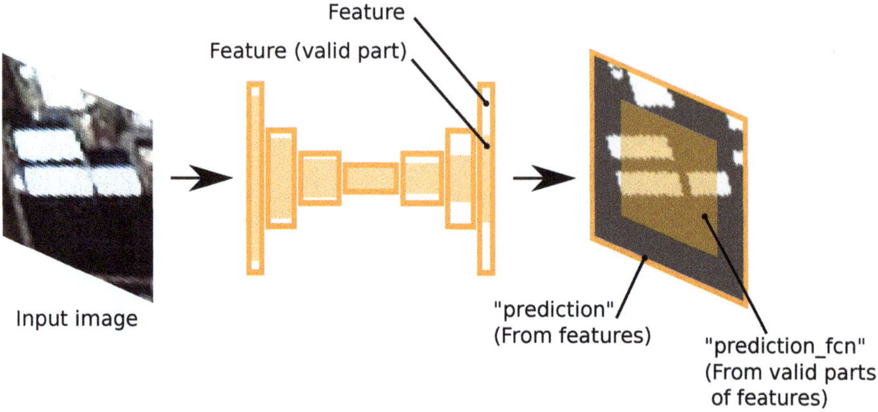

FIGURE 12.10
At each layer, the feature maps are contaminated by the zero padding of the convolutions. Blocks with orange borders: features; blocks filled in orange: valid parts of feature maps. The *exact* part of the final output feature map is its center area.

Table 12.1 shows how to compute the number of pixels to remove from the border of the last feature map (the *prediction* output tensor). At each layer, convolution contaminates the feature map due to padding. We can calculate the central part of the feature maps, which is *exact*, i.e. not contaminated by padding. As all convolution kernels are size 3, it's easy to compute the number of pixels that should be clipped at the end. For instance: Level 1, convolution output is 32 (input is 64, and we use convolutions with stride 2) but as our kernel is 3×3, the actual valid part of the output feature map has a size of 31×31.

TABLE 12.1
Valid part of the convolution outputs (namely feature maps)

Level	Feature size	Valid part
/	64	64
1	32	31
2	16	15
3	8	7
4	4	3
1	8	5
2	16	9
3	32	17
4	64	33

As shown in table 12.1, the valid part of the final feature map has a size 33×33 elements. We then remove 16 elements from the borders to output a valid tensor of 32×32 elements.

12.2.2.2 Expression field

In OTBTF the spatial part that the model outputs for the tensors specified in the **output.names** is called the *expression field.* As explained, the model transforms an elementary input image of size 64×64 into an elementary output predicted label image of size 32×32. Hence, later we will use a receptive field of 64×64 (the input volume that the network "sees") and an expression field of 32×32 (the output volume that the network "fills").

12.2.3 Generate the SavedModel

To generate the model, use the *create_model4.py* Python script.

```
# Generate the model

# The "create_model4.py" file can be downloaded from:
# https://github.com/remicres/otbtf_tutorials_resources.git
python3 ~/python/create_model4.py "model4"
```

Important note
As you can note, we carefully set 4 dimensions for the output tensors that will be used in the OTBTF applications. Currently, OTBTF interprets the last dimension of an output tensor as the number of components of the output image. That is why we use *tf.expand_dims()* to make the *prediction* output tensor with a last dimension of size 1.

12.3 Training the model

Train the model with the **TensorflowModelTrain** application:

```
# Since the TensorflowModelTrain uses n+1 sources
export OTB_TF_NSOURCES=1

# Train the deep learning model
otbcli_TensorflowModelTrain \
-training.source1.il amsterdam_Spot7_patches_A.tif \
-training.source1.patchsizex 64 \
-training.source1.patchsizey 64 \
-training.source1.placeholder "x" \
-training.source2.il amsterdam_labelimage_patches_A.tif \
-training.source2.patchsizex 64 \
-training.source2.patchsizey 64 \
-training.source2.placeholder "y" \
-model.dir model4 \
-training.targetnodes optimizer \
-validation.mode "class" \
-validation.source1.il amsterdam_Spot7_patches_B.tif \
-validation.source1.name "x" \
-validation.source2.il amsterdam_labelimage_patches_B.tif \
-validation.source2.name "prediction" \
-model.saveto model4/variables/variables
```

12.4 Inference

As explained in section 12.2.1, we set an input receptive field of size 64×64 and an output expression field of size 32×32. We ask the **TensorflowModelServe** application to produce, in fully convolutional mode, the *prediction_fcn* output tensor of the model.

Let's run the model over a subset image:

```
# Produce the map
otbcli_TensorflowModelServe \
-source1.il IMG_SPOT7_amsterdam_PMS_norm.tif \
-source1.rfieldx 64 \
-source1.rfieldy 64 \
-source1.placeholder "x" \
-model.dir model4 \
-model.fullyconv on \
-output.efieldx 32 \
-output.efieldy 32 \
-output.names "prediction_fcn" \
-out "classif_model4.tif?&box=6000:4000:2000:2000" uint8
```

Ask the **TensorflowModelServe** application to produce larger blocks (tweaking the *optim.tilesizex* and *optim.tilesizey* parameters to speed up the process) and produce the entire buildings map. Open the output in QGIS (ref to figure 12.11).

```
# Produce the map
otbcli_TensorflowModelServe \
-source1.il IMG_SPOT7_amsterdam_PMS_norm.tif \
-source1.rfieldx 64 \
-source1.rfieldy 64 \
-source1.placeholder "x" \
-model.dir model4 \
-model.fullyconv on \
-output.efieldx 32 \
-output.efieldy 32 \
-output.names "prediction_fcn" \
-optim.tilesizex 2048 \
-optim.tilesizey 2048 \
-out classif_model4_full.tif uint8
```

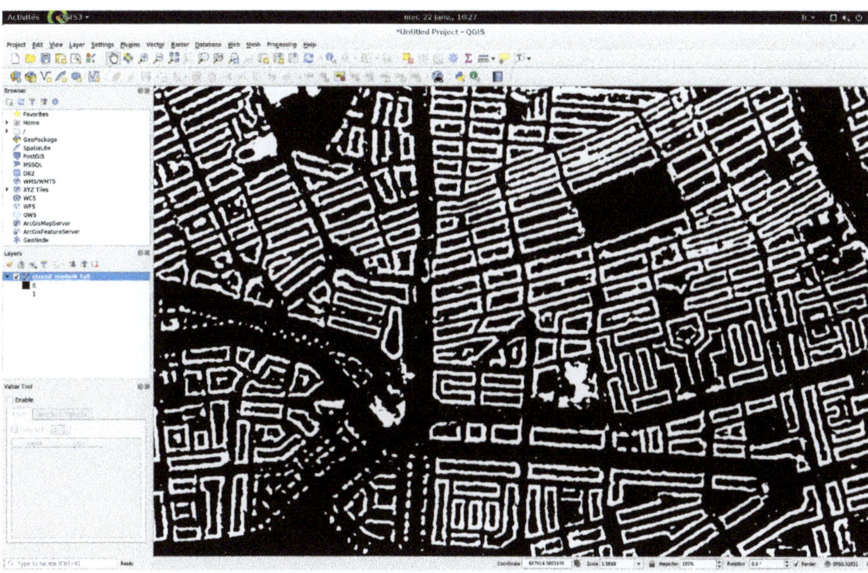

FIGURE 12.11
Import the output classification map in QGIS.

Note on evaluation metrics
The usual metrics to evaluate a classification map are the precision, f-score and accuracy for each class, and overall kappa index and accuracy

index. Regarding these metrics values (after training the network for semantic segmentation), there are two things to notice:

1. The metrics values of class 0 (*background*) are high. This comes from the fact that classes are unbalanced: there are many more pixels from the background as for the other classes.
2. One should carefully interpret these evaluation metrics. The kappa index and overall accuracy index are not suited for unbalanced classes or precision of classes. The F-score might be preferred for land cover map evaluation since it expresses a compromise between precision and recall.

13

Discussion

This part provided insight into semantic segmentation on remote sensing images. We have introduced semantic segmentation using a U-Net-like architecture, a popular model targeting semantic segmentation of images.

We have built and used a small model to classify the buildings over one entire remote sensing image, without any limitation regarding its size, and in reasonable time.

We saw that, unlike the patch-based approach, semantic segmentation employs models that preserve the spatial resolution of the output. However, we have also understood that this kind of model must be trained from densely annotated images: the training step requires patches of images that are fully annotated, i.e. each pixel of the image has a corresponding class label. This complicates the elaboration of reference data, however more and more open data are available (e.g. OpenStreetMap).

We presented a simple framework to prepare, select and extract those patches in the remote sensing images, and from a rasterized OpenStreetMap vector layer of building footprints.

We built a small modified U-Net model that inputs a pansharpened Spot image, and generates a land cover map. Thanks to the so-called skip-connections, that reuse the features computed from different scales within the network, our model has preserved the semantic details. An interesting extension of this exercise could be to skip the pansharpening step, and directly fuse the multispectral channels in the network. This could be realized with a two-branch encoder that separately inputs the panchromatic channel at 1.5 m and the multispectral image at 6.0 m. The two feature maps obtained at the same scale could then be concatenated. For instance, after two strided convolutions, feature maps of the panchromatic encoder branch would have the same physical spacing as the multispectral image (or its feature maps, if no stride in the multispectral encoder branch) and thus could be concatenated together.

We also presented a method to properly use fully convolutional models on very large images, avoiding blocking artifacts: we have extracted the central area of the model output, that represents the valid result of the convolutions, to compute in a fully convolutional fashion large regions of the input image. Among other benefits of this approach are reproducibility (one can process an entire image, or a subset: result will be identical) and speed.

It should be noted that deep learning also offers plenty of architectures for semantic segmentation, and each year new models emerge with various different strategies. To name a few, we can cite DeepLab [26], which uses the atrous convolutions, or PSPNet [27], which uses a Pyramid Pooling Module.

Deep learning is a growing trend, and each year new approaches appear. We have no doubt that in the near future, plenty of new methods for semantic segmentation will appear, leveraging the remaining difficulties ahead, semantic segmentation model training, and particularly regarding data availability.

Part IV

Image restoration

14

Gapfilling of optical images: principle

In this section, we propose to address a problem of image restoration, which is quite different in nature than land cover mapping, to show that deep learning can also be applied to other topics in remote sensing. For the practice session, using a GPU is strongly advised to reduce the processing time. We refer the reader to the OTBTF GitHub repository[1] to get instructions for using the Docker image with GPU support.

14.1 Introduction

A well-known issue impacting optical imagery is the presence of clouds. The need for cloud-free images at precise date is required in many operational monitoring applications. On the other hand, the SAR sensors are cloud-insensitive and they provide orthogonal information with respect to optical satellites that enable the retrieval of information, which could be lost in optical images because of cloud cover. To alleviate this issue, a common approach consists of interpolating temporally close optical images to approximate the missing contents of the target image [28], but this method uses only optical images, and is not designed to exploit the SAR information. Other machine learning–based approaches such as dictionary learning [29, 30], and more recently deep learning [31, 32], have emerged to restore cloud-impacted optical images with SAR images. For instance in [33], optical and SAR time series are jointly processed with a CNN to retrieve the Normalized Difference Vegetation Index of the missing optical image.

14.2 Method

In this exercise, we apply the method described in [34]. This method consists of reconstructing (gapfilling) one optical image, polluted by cloud phenomena,

[1] `https://github.com/remicres/otbtf`

exploiting available multi-temporal optical images, and one SAR image which is acquired in the same time as the polluted image. The approach aims to restore gaps in the optical image to provide a cloud-free product, and has shown superior results compared to traditional gapfilling (i.e. linear interpolation or linear regression) in terms of root mean squared error. Given t, the time stamp of the optical image polluted by clouds (denoted $S2_t$), the method uses one SAR image acquired at t (denoted $S1_t$), and a couple of available cloud-free optical images acquired before and after t (respectively denoted $S2_{t-1}$ and $S2_{t+1}$), to estimate a synthetic optical image at t (denoted $\widehat{S2_t}$) that can be used to fill the missing values of the actual optical image $S2_t$. We intend to estimate $\widehat{S2_t}$ from $S2_{t-1}$, $S1_t$, $S2_{t+1}$ using a deep neural network called a *deep convolutional autoencoder*, which is trained over the available parts of the images (illustrated in figure 14.1). We focus on the minimization of the spread between the cloud-free parts of the target image (i.e. the ground truth), and the estimated image generated from our deep network. The training of our deep network hence consists of minimizing a cost function that evaluates the distance between $\widehat{S2_t}$ and $S2_t$ images in terms of L_2 loss. We assume that the SAR image and the optical images cover the same physical extent and are superimposed over the same physical pixel grid, meaning they share the same coordinate reference system, origin, physical pixel spacing, and size. We also assume that optical images have the same number of components per pixel (i.e. same number of spectral bands). In the following, we will use these notations for the images:

- $S2_{t-1}$ for the optical image acquired at date $t-1$
- $S1_t$ for the SAR image acquired at a date close to t
- $S2_{t+1}$ for the optical image acquired at date $t+1$
- $\widehat{S2_t}$ for the estimated optical image acquired at date t
- $S2_t$ for the actual optical image acquired at the date t

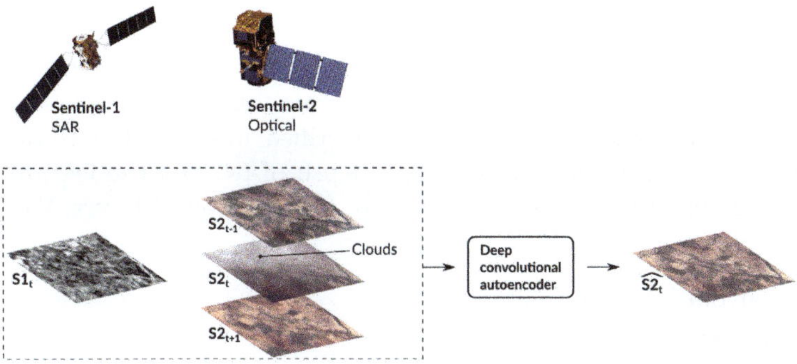

FIGURE 14.1
Our deep-learning-based approach for optical image gapfilling. The deep convolutional autoencoder is trained with every available source (the optical images $S2_{t-1}$, $S2_t$, $S2_{t+1}$ and the SAR image $S1_t$), for the purpose of estimating the $\widehat{S2_t}$ image.

14.3 Architecture

Our network inherits from the U-Net architecture described in [25]. As we mentioned in chapter 10, U-Net is a popular model used in semantic labeling or image-to-image translation. Our architecture is presented in figure 14.2. The main differences between our architecture and the original U-Net, is that we have three inputs instead of one, hence three encoder branches. Each encoder branch of the model encodes, respectively, the input images $S2_{t-1}$, $S1_t$ and $S2_{t+1}$ in to a feature map called *the bottleneck*. This bottleneck is the concatenation of the last feature maps of the encoders. The second part of the network decodes this bottleneck to retrieve the estimated image $\widehat{S2_t}$.

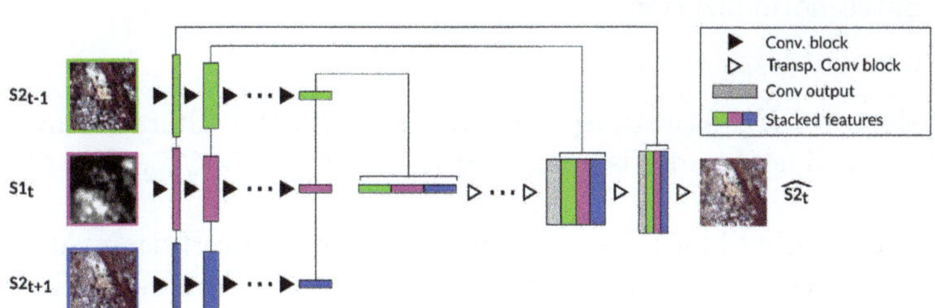

FIGURE 14.2
Architecture used for optical image gapfilling.

14.3.1 Encoder

Each encoder implements a convolutional neural network with strides. It consists of successive 4×4 convolutions with stride 2 to downsample the input. Each convolution is followed by a leaky rectified linear unit (LReLU). The first outputs a feature map which has a depth of 64, and the feature map's depth is doubled at each convolution except for the two last layers. We denote C_k as a block composed of the following:

1. Convolution with k filters
2. LReLU layer
3. Batch normalization layer [35]
4. Dropout layer

Each encoder is composed as follows: C_{64}-C_{128}-C_{256}-C_{512}-C_{512}-C_{512}.

14.3.2 Decoder

The decoder is built as the mirror of the encoder. It consists of successive 4×4 transposed convolutions with stride 2 to upsample the input. The first transposed convolution inputs the 512-element-depth bottleneck feature maps. The depth of the feature maps is divided by 2 at each transposed convolution. Each transposed convolution is followed by a rectified linear unit (ReLU). It is then concatenated with the corresponding feature maps from the encoder. The main difference between our architecture and the original U-Net is that it has three encoder branches that are stacked together as the bottleneck. The last transposed convolution outputs a 4-element-depth feature map, which corresponds to the number of channels of the Sentinel-2 image, followed by a hyperbolic tangent unit. We denote C_k^{tr} as a block containing the following:

1. A transposed convolution with k filters
2. ReLU
3. Batch normalization
4. Dropout layer

We also denote CH_k^{tr} as a transposed convolution with k filters + a hyperbolic tangent layer. The decoder is composed as follows: C_{512}^{tr}-C_{512}^{tr}-C_{256}^{tr}-C_{128}^{tr}-C_{64}^{tr}-CH_4^{tr}.

In most building blocks, we use dropout to avoid overfitting. As our network has a total of 6 levels of convolutions/transposed convolutions, it is suited to input 64×64 input patches of $(S2_{t-1}, S1_t, S2_{t+1})$ to estimate 64×64 output patches of $\widehat{S2_t}$.

14.3.3 Loss

We train the network parameters from a training set T composed of n tuples t_i of $(S2_{t-1},\ S1_t,\ S2_{t+1},\ S2_t)$ patches.

$$T = \{1, ..., t_n\}\,, t_i = (S2_{t-1,i}\,, S1_{t,i}\,, S2_{t+1,i}\,, S2_{t,i}\,) \tag{14.1}$$

The objective of the training phase is to find:

$$\arg\, \min \mathbb{E}(L_2(\widehat{S2_t}, S2_t)) \tag{14.2}$$

We use the L_2-norm since it expresses the Euclidean distance between pixels of the real image and the estimated image. Hence, the average of L over all patches corresponds to the root mean squared error of the estimation. To optimize our network, we perform the gradient descent of L using the Adam solver [19].

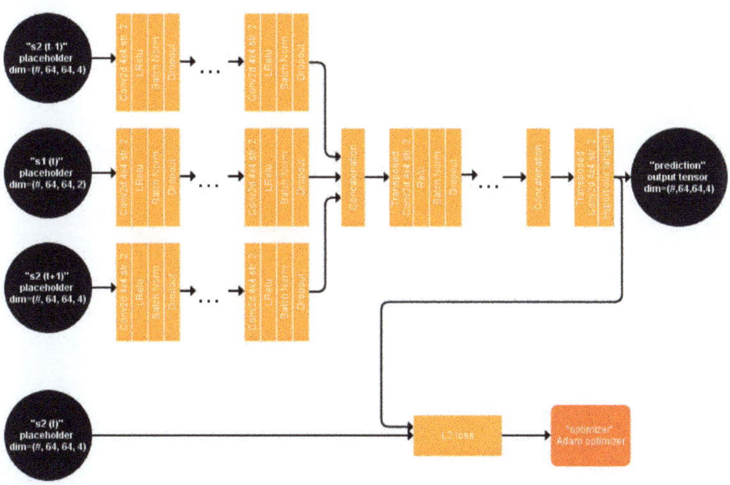

FIGURE 14.3
Model implementation. For the sake of clarity, the concatenation of encoder and decoder feature maps are not represented. The dots represent the rest of the 6 building blocks of the encoder and the decoder.

As in section 5.3, we give the name *optimizer* to the operator for the Adam solver. Figure 14.3 resumes the model training implementation.

15

The Marmande dataset

This section aims to present the data involved in this practice session. We rely on freely available data, named the *Marmande* dataset. In the following chapter, we explain how to download and prepare this dataset in order to be ready for the practice session. Create a directory named *marmande* on your file system. We will use this directory to store all the downloaded and generated data during the exercises.

15.1 Description

The *Marmande* dataset consists of three Sentinel-2 images and one Sentinel-1 product. It will be used for all exercises of section IV. Table 15.1 summarizes the image characteristics.

TABLE 15.1
Characteristics of the images

Sensor	Type	*col* × *row* × *band*	Spacing	Acquisition	EPSG
S2	Optical	10980 × 10980 × 4	10m	2019/07/05	32630
S2	Optical	10980 × 10980 × 4	10m	2019/07/25	32630
S2	Optical	10980 × 10980 × 4	10m	2019/08/24	32630
S1	SAR	30010 × 15616 × 2	10m	2019/07/25	4326

15.2 Sentinel-2 images

We use Sentinel images acquired over the France mainland, in the area of Marmande, which is close to the city of Bordeaux (figure 15.1).

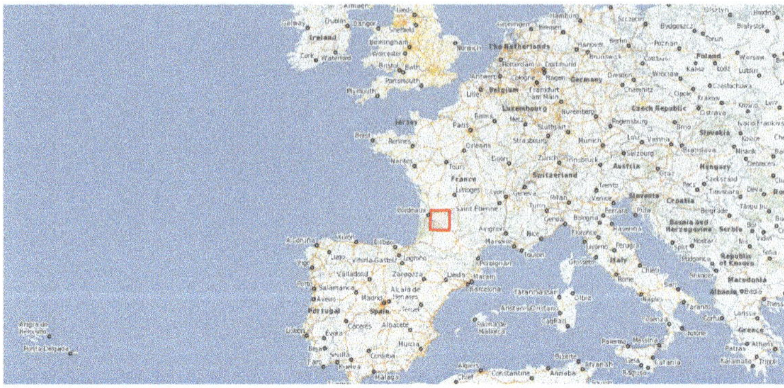

FIGURE 15.1
The Region of Interest of the Marmande dataset.

In the following, we detail how to download the products. As in section 4, we will use the European Space Agency Hub to download the Sentinel-2 images acquired over the **T30TYQ** tile, on the following dates:

- 2019/07/05, as $S2_{t-1}$
- 2019/07/25, as $S2_t$
- 2019/08/24, as $S2_{t+1}$

In the following, we describe the steps to download the products. First, log in to the European Space Agency hub.[1] Then, search the products having the following attributes:

- Area: Marmande (You can zoom over the area of Marmande, as shown on figure 15.2, and draw a region of interest on the map.)
- Sensing date: for instance, for $S2_{t-1}$: from *2019/07/05* to *2019/07/05*
- Mission: tick the *Sentinel-2* box
- Product type: *S2MSI2A*

[1] `https://scihub.copernicus.eu/dhus/#/home`

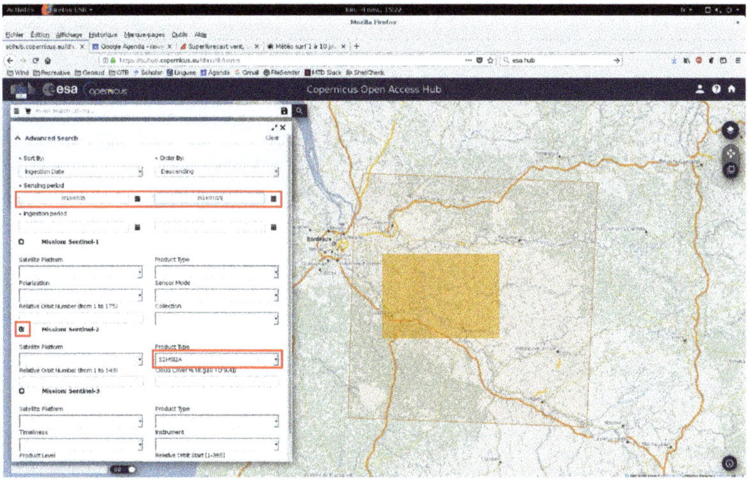

FIGURE 15.2
Enter the following attributes to search the Sentinel-2 product: Mission: *Sentinel-2*, Product type: *S2MSI2A*. Then select the area of Marmande, enter the Sensing date and validate the search for each image to download.

Repeat this step for each product. You should be able to download these three products in your *marmande* directory:

1. S2A_MSIL2A_20190705T105031_N0212_R051_T30TYQ_20190705T140734

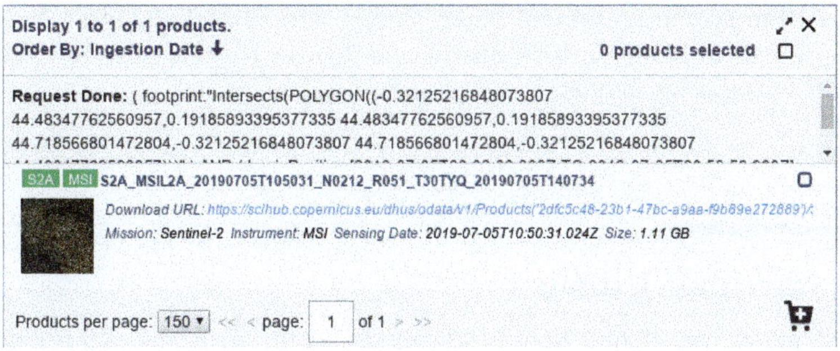

2. S2A_MSIL2A_20190725T105031_N0213_R051_T30TYQ_20190725T141902

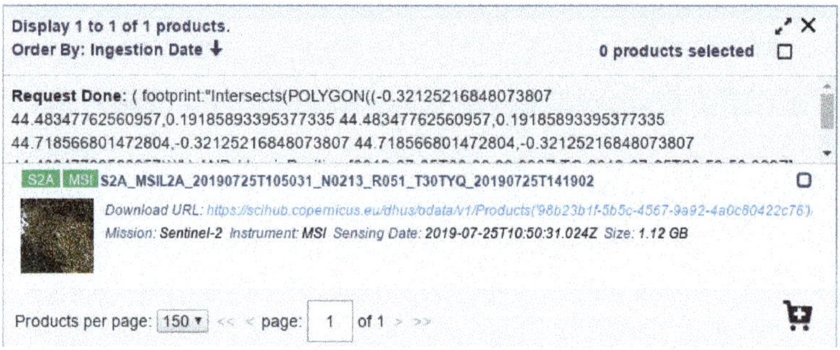

3. S2A_MSIL2A_20190824T105031_N0213_R051_T30TYQ_20190824T134703

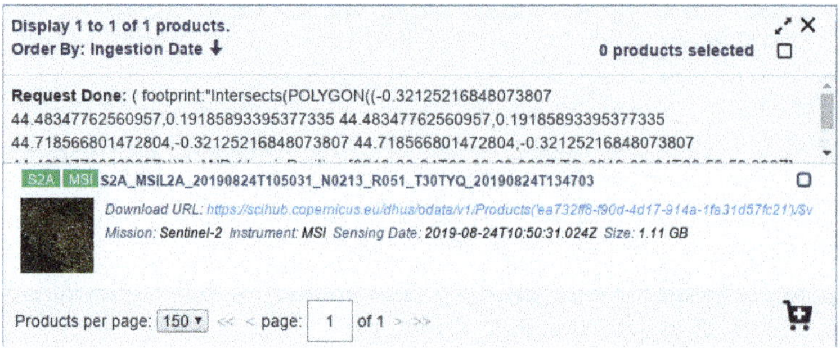

Once the products have been downloaded, you can unzip the archives and look into the extracted files. The Sentinel-2 product has two sets of images: one has 10-m spacing (spectral bands 2, 3, 4 and 8) and the other has 20-m spacing (spectral bands 5, 6, 7, 8a, 11 and 12). In this exercise, we will only use the 10-m-spacing bands.

15.3 Sentinel-1 image

We will make use of one Sentinel-1 image acquired in the same region of interest, as shown in figure 15.1, and most importantly, on the same exact date as the Sentinel-2 image that we want to reconstruct, $S2_t$. Here we use again the European Space Agency hub to download the product.

1. Draw a rectangle corresponding roughly to the Marmande Region of Interest (see figure 15.1).
2. Open the search tab.
3. Set the *Sensing period* from 2019/07/25 to 2019/07/25 to filter only products that have been acquired on the same date as $S2_t$.

4. Tick the *Mission: Sentinel-1* box.
5. In *Product type*, select *GRD*.
6. In *Polarization*, select *VV+VH*.
7. In *Sensor mode*, select *IW*.
8. In *Satellite Platform*, select *S1A_**.
9. Click on the search button. The list of matching products should appear, and display a single product named *S1A_IW_GRDH_1SDV_20190725T175553_20190725T175618_028277_0331C8_3C98*.

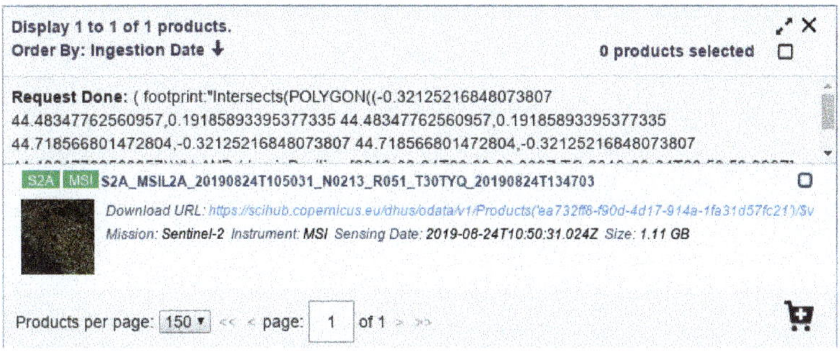

16

Pre-processing

The Sentinel-1 and Sentinel-2 images should now be downloaded and ready to use. We will proceed to the creation of a dataset that can be used (i) to train the model described in section 14.3, and also (ii) to evaluate its performance in terms of root mean squared error.

16.1 Sentinel images

The Sentinel-1 image has 16 bits signed encoding and the Sentinel-2 images have 16 bits unsigned encoding, i.e. integer numbers. We need to normalize the pixels to have absolute values inside the $[0, 1]$ range. As in sections 5.1, 8.3 and 12.1.2 we use the **BandMathX** application of the Orfeo ToolBox to perform the band concatenation and the normalization of values.

16.1.1 Optical images

We process only the 10-m-resolution bands, numbers 2, 3, 4 and 8. The actual encoding of the Sentinel-2 sensor is 12 bits. In Sentinel images, the 16-bit encoding is used for convenience, because the Jpeg2000 driver, as most remote sensing images drivers, doesn't support 12-bit encoding. But the actual digital number produced by the Sentinel-2 ground segment has 12-bit encoding. Hence, a simple method to perform the normalization of input images consists of dividing the pixel values by the maximum value of 12 bits, i.e. 8192. We can use **BandMathX** to apply the corresponding linear transformation to the pixel values:

$$x_{norm} = \alpha x \tag{16.1}$$

With:

$$\begin{aligned} \alpha &= \frac{1}{2^{12}} \\ &= \frac{1}{8192} \\ &= 1.2207 \times 10^{-4} \end{aligned} \tag{16.2}$$

Enter the following expression in **BandMathX** to process the first image of the Sentinel-2 time series:

```
# Normalize the Sentinel-2 image from 2019/07/05

# Go in Sentinel-2 images (10m)
cd ~/marmande
cd S2A_MSIL2A_20190705T105031_N0212_R051_T30TYQ_20190705T140734.SAFE
cd GRANULE/L2A_T30TYQ_A021069_20190705T105032/IMG_DATA/R10m

# Concatenate images, and normalise them between [0,1]
otbcli_BandMathX \
-il T30TYQ_20190705T105031_B04_10m.jp2 \
    T30TYQ_20190705T105031_B03_10m.jp2 \
    T30TYQ_20190705T105031_B02_10m.jp2 \
    T30TYQ_20190705T105031_B08_10m.jp2 \
-exp "{0.00012207*im1b1;0.00012207*im2b1;␣\
0.00012207*im3b1;0.00012207*im4b1}" \
-out ~/marmande/Sentinel2_2019-07-05_10m.tif

# Go back in the workspace
cd ~/marmande
```

Repeat the operation on the two next Sentinel-2 images. First, apply **BandMathX** to the *S2A_MSIL2A_20190725T105031_N0213_R051_T30TYQ_20190725T141902* product, which corresponds to $S2_t$:

```
# Normalize the Sentinel-2 image from 2019/07/25
cd ~/marmande
cd S2A_MSIL2A_20190725T105031_N0213_R051_T30TYQ_20190725T141902.SAFE
cd GRANULE/L2A_T30TYQ_A021355_20190725T105702/IMG_DATA/R10m
otbcli_BandMathX \
-il T30TYQ_20190725T105031_B04_10m.jp2 \
    T30TYQ_20190725T105031_B03_10m.jp2 \
    T30TYQ_20190725T105031_B02_10m.jp2 \
    T30TYQ_20190725T105031_B08_10m.jp2 \
-exp "{0.00012207*im1b1;0.00012207*im2b1;␣\
0.00012207*im3b1;0.00012207*im4b1}" \
-out ~/marmande/Sentinel2_2019-07-25_10m.tif
cd ~/marmande
```

Then, apply it to the *S2A_MSIL2A_20190824T105031_N0213_R051_T30TYQ_20190824T134703* product, which corresponds to $S2_{t+1}$:

```
# Normalize the Sentinel-2 image from 2019/08/24
cd ~/marmande
cd S2A_MSIL2A_20190824T105031_N0213_R051_T30TYQ_20190824T134703.SAFE
cd GRANULE/L2A_T30TYQ_A021784_20190824T105344/IMG_DATA/R10m
otbcli_BandMathX \
-il T30TYQ_20190824T105031_B04_10m.jp2 \
    T30TYQ_20190824T105031_B03_10m.jp2 \
    T30TYQ_20190824T105031_B02_10m.jp2 \
    T30TYQ_20190824T105031_B08_10m.jp2 \
-exp "{0.00012207*im1b1;0.00012207*im2b1;␣\
0.00012207*im3b1;0.00012207*im4b1}" \
-out ~/marmande/Sentinel2_2019-08-24_10m.tif
cd ~/marmande
```

At this point you can import in QGIS the three generated images *Sentinel2_2019-07-05_10m.tif* ($S2_{t-1}$), *Sentinel2_2019-07-25_10m.tif* ($S2_t$), and *Sentinel2_2019-08-24_10m.tif* ($S2_{t+1}$). QGIS will display the images in true colors since the index of the red, green and blue channels of the pre-processed Sentinel-2 images are respectively 1, 2 and 3.

16.1.2 SAR image

Radar is an active sensor, meaning that it emits a radio wave "illuminating" the target (but not in the visible domain) and the echo is received and recorded. Synthetic Aperture Radar (SAR) is a kind of radar that generates images using the motion of the sensor to simulate a larger antenna, enabling a higher-resolution image. The acquisition principle is that the target is viewed multiple times from the moving sensor, but with different acquisition angles. SAR processing combines these acquisitions and allows us to increase the resolution of the original antenna. A few pre-processing steps are required prior to processing SAR images.

16.1.2.1 Calibration

The first step to prepare the SAR images is the calibration. The calibration ensures that the intensity value actually represents the sigma zero (σ_0) value of the reflectivity, i.e., the radar cross section normalized to area. We use the SARCalibration application of the Orfeo ToolBox to compute the calibrated σ_0 image from the downloaded Sentinel-1 product.

We first generate the calibrated VH image.

```
# SAR calibration
cd ~/marmande
cd S1A_IW_GRDH_1SDV_20190725T175553_20190725T175618_028277_0331C8_3C98
.SAFE
cd measurement

# Calibration of VH channel
otbcli_SARCalibration \
-in s1a-iw-grd-vh-20190725t175553-20190725t175618-028277-0331c8-002
.tiff \
-out ~/marmande/sar_vh.tif

# Go back to workspace
cd ~/marmande
```

We repeat the same operation to generate the VV image.

```
# SAR calibration
cd ~/marmande
cd S1A_IW_GRDH_1SDV_20190725T175553_20190725T175618_028277_0331C8_3C98
.SAFE
cd measurement

# Calibration of VV channel
```

```
otbcli_SARCalibration \
-in s1a-iw-grd-vv-20190725t175553-20190725t175618-028277-0331c8-001
.tiff \
-out ~/marmande/sar_vv.tif

# Go back to workspace
cd ~/marmande
```

16.1.2.2 Filtering values

SAR images are now calibrated. The physical measurement unit of the generated images is a linear ratio between the emitted and the received energy of the radio wave. A particular effect to be observed in SAR images is the *speckle*, which is inherent in the imaging. During the calibration step, the average noise is removed. However, the SAR speckle noise is still present in the calibrated images in the form of extreme values. In the following step, we use **BandMathX** to remove the extremely high values. In particular, we clip all pixels values x where $x > 1$: the physical meaning of such values is that the sensor received more energy than it emitted.

```
# SAR clipping in dB scale
otbcli_BandMathX -il sar_vh.tif sar_vv.tif \
-exp "vlog({min(max(im1b1,0.00001),1);min(max(im2b1,0.00001),1)})" \
-out sar_vh_hh_log.tif
```

16.1.2.3 Linear stretch

Finally, we transform the pixel values, using a linear function to map the absolute pixel values in the $[0, 1]$ range.

```
# Normalize
otbcli_BandMathX -il sar_vh_hh_log.tif \
-exp "{im1b1/im1b1Mini;␣im1b2/im1b2Mini}" \
-out sar_vh_hh_log_norm.tif
```

16.1.2.4 Spatial resampling

Finally, we need to reproject the image in the same coordinate reference system as the Sentinel-2 images, i.e. EPSG:32630. We also want to resample the image to the same extent, and with the same pixel spacing as the Sentinel-2 image: this way all Sentinel-1 and Sentinel-2 images have the same spatial extent/size/coordinate reference system and patches can be extracted trivially. To perform this geometric operation, we use the Orfeo ToolBox application named **Superimpose**. This application performs resampling of an input image (the *moving image*) into the same grid/reference system as a target image (the *reference image*). These following parameters are used in the Superimpose application:

- **inr** is the *reference image*. We use one of the Sentinel-2 images, for instance the *Sentinel2_2019-07-05_10m.tif* image.

- **inm** is the *moving image*. In our case, the image to resample is the Sentinel-1 image we have already radiometrically pre-processed, the *sar_vh_hh_log_norm.tif* image.

- **out** is the output image, that is, the *moving image* resampled to the same spatial extent and with the same pixel spacing as the *reference image.*

Perform the resampling with the following command line.

```
# Resample the Sentinel-1 image over one Sentinel-2 image
otbcli_Superimpose \
-inm sar_vh_hh_log_norm.tif \
-inr Sentinel2_2019-07-05_10m.tif \
-out sar_vh_hh_log_norm_EPSG32630.tif
```

16.2 Patches

The Sentinel-1 and Sentinel-2 images are pre-processed, and we can now proceed to the next step: the preparation of patches. The goal is to prepare two datasets: one for training the network (the training dataset), and the other to evaluate the performance of the network (the validation dataset). Each dataset consists of quadruplets of ($S2_{t-1}$, $S2_t$, $S2_{t+1}$, $S1_t$) patches.

16.2.1 Patch position seeding

The first step is to select the locations where the samples will be extracted. Here, we must select some clean patches in the images: patches must be cloud-free and have no-data values. To accomplish this task, we will use QGIS exclusively in performing these steps:

1. For each Sentinel-2 image, we create a mask of pixels that are not valid (i.e. cloudy or *no-data* values). Since the Sentinel-1 image is cloud-free and entirely overlaps our region of interest, we don't need to create a mask for it.
2. We then merge the Sentinel-2 masks to obtain a unique mask of the areas that should be avoided in extracting the samples.
3. Finally, we generate a grid of points, and keep the points that lie in the valid part of the Sentinel-2 images, i.e. which do not intersect the mask of areas to avoid. We use a grid because (i) the samples are uniformly distributed in space, and (ii) the patches do not overlap.

16.2.1.1 Sentinel-2 image masks

In this part, we will create one mask of polluted/*no-data* pixels for each Sentinel-2 image. We detail the steps to create the mask of one Sentinel-2 image, *Sentinel2_2019-07-05_10m.tif.* You will then be able to process the two other Sentinel-2 images, *Sentinel2_2019-07-25_10m.tif* and *Sentinel2_2019-08-24_10m.tif*, by repeating the same operations.

Create a new empty project in QGIS, and import the *Sentinel2_2019-07-05_10m.tif* image.

The objective is to create a vector layer representing the area where pixels of the *Sentinel2_2019-07-05_10m.tif* image are cloudy or have *no-data* values. For the sake of simplicity, we manually draw polygons that represent these areas.

First, create a new vector layer: from the menu, select **Layer** > **Create Layer** > **New geopackage layer** as shown in figure 16.1.

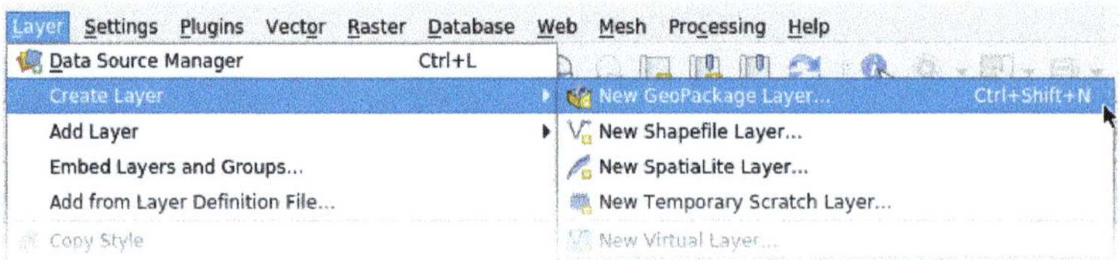

FIGURE 16.1
Create a new vector layer from the menu.

In *Geometry type*, select *Polygon* and save the vector layer in a file named *Sentinel2_2019-07-05_holes.gpkg* in your working directory, as shown in figure 16.2.

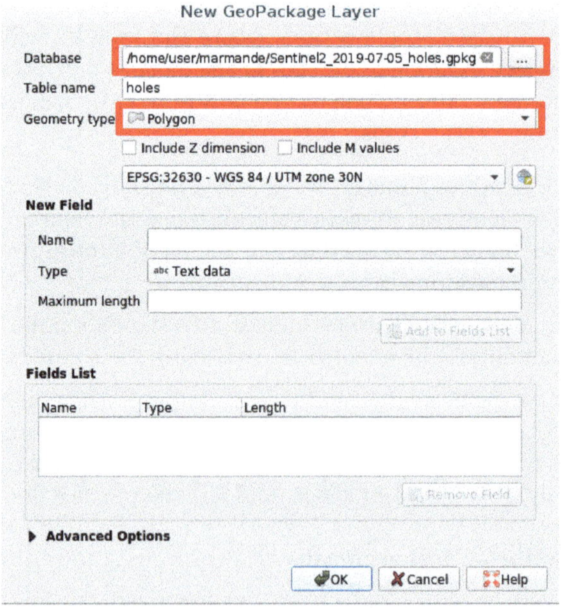

FIGURE 16.2
Save the vector layer into *Sentinel2_2019-07-05_holes.gpkg*.

Now draw polygons that include the clouds and the *no-data* values from the image *Sentinel2_2019-07-05_10m.tif*. First, you need to toggle editing mode as shown in figure 16.3.

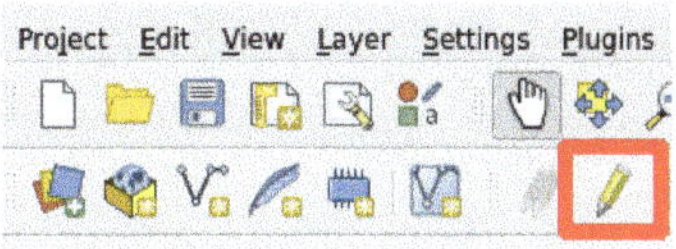

FIGURE 16.3
The *Toggle editing* button.

To draw a new polygon, click on the *Add polygon features* button as shown in figure 16.4.

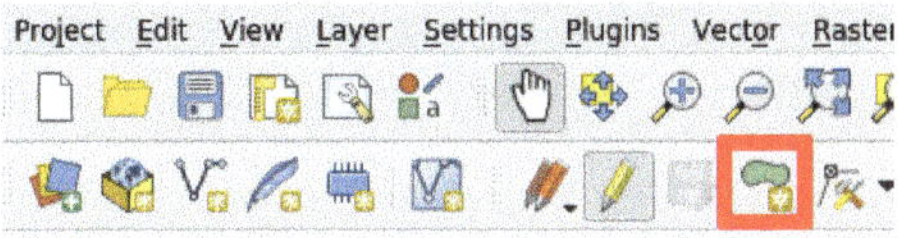

FIGURE 16.4
The *Add polygon features* button.

You can now draw the polygons around areas to reject, and save the modifications by clicking again on the editing button (figure 16.3). There are two kinds of areas to reject: the clouds and the borders of the image, where no data has been captured by the sensor and the result is "black" pixels (denoting the *no-data* values). Create polygons around these areas and repeat the steps to create one vector layer per Sentinel-2 image. In the end, your masks should look like those illustrated in figure 16.5.

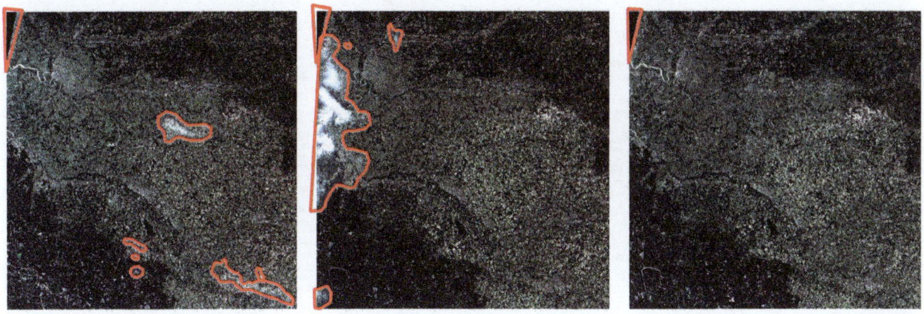

FIGURE 16.5
Vector layers (.gpkg) of cloudy/no-data pixels are in red. Left: *Sentinel2_2019-07-05_10m.tif* and *Sentinel2_2019-07-05_holes.gpkg*, center: *Sentinel2_2019-07-25_10m.tif* and *Sentinel2_2019-07-25_holes.gpkg*, right: *Sentinel2_2019-08-24_10m.tif* and *Sentinel2_2019-08-24_holes.gpkg*.

16.2.1.2 Merge masks

You should have ended the previous step with 3 new vector layers:

1. *Sentinel2_2019-07-05_holes.gpkg*
2. *Sentinel2_2019-07-25_holes.gpkg*
3. *Sentinel2_2019-08-24_holes.gpkg*

Since we want samples that are valid in the four images ($S2_{t-1}$, $S2_t$, $S2_{t+1}$ and $S1_t$), we need to merge the vector layers of images into a single one that indicates where at least one image has missing content. Use QGIS to merge the masks into a single vector layer. From the menu, select **Vector** > **Data management tools** > **Merge vector layers** ref to figure 16.6.

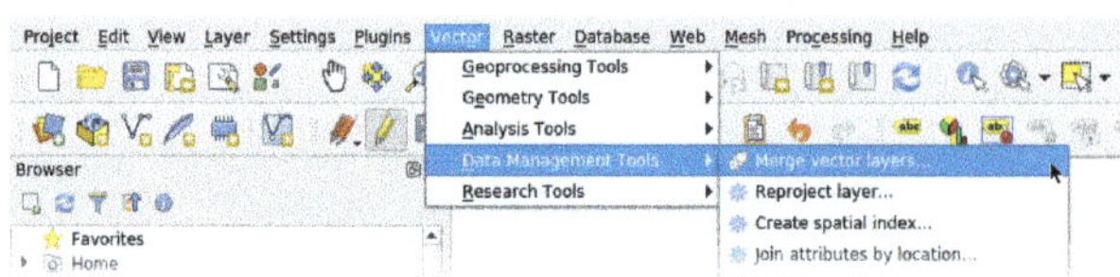

FIGURE 16.6
Merge the masks.

In *input layers*, select the 3 vector layers *Sentinel2_2019-07-05_holes.gpkg*, *Sentinel2_2019-07-25_holes.gpkg* and *Sentinel2_2019-08-24_holes.gpkg*. Select the destination CRS (Coordinate Reference System) as EPSG:32630, which is the same as the Sentinel images. Save the output vector layer as a temporary layer. Figure 16.7 summarizes the parameters to provide.

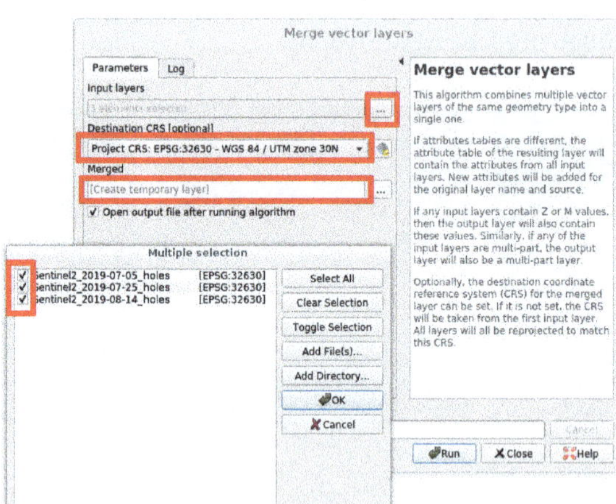

FIGURE 16.7
Select the masks to merge. Keep the same coordinate reference system as the images (EPSG:32630). Save the output vector layer as a temporary layer.

The merge mask should look like the illustration in figure 16.8.

FIGURE 16.8
The resulting *Merged* layer.

16.2.1.3 Grid generation

In this section, we will generate the grid that will be used to extract the patches in the images. As in section 12.1.3.1, we use QGIS to generate a vector layer of polygons that represent the patch layout, aligned on the pixel positions of the images. Since our model needs 64 × 64 patches at inputs and the pixels have 10-m spacing, we will generate a grid tiled with 640-meter-sized squares. From the menu, click on **Vector** > **Research tools** > **Create grid** as shown in figure 16.9.

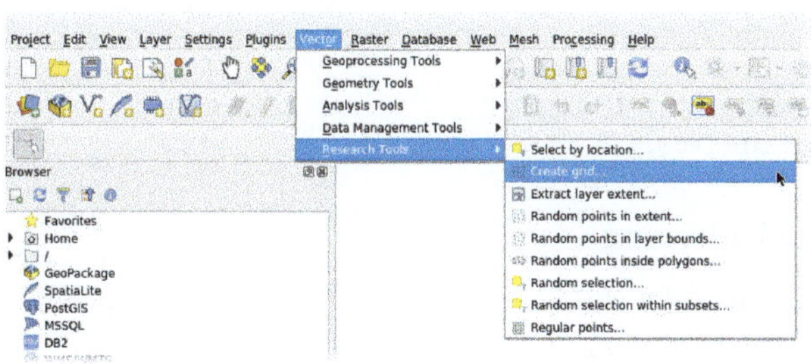

FIGURE 16.9
Create a vector grid.

Once the **Create grid** tool is open, set the following fields (figure 16.10):

1. **Grid type**: select *Rectangle*.
2. **Grid extend**: set *Use layer extent*, then select one of the images, for instance, *Sentinel2_2019-07-05_10m.tif*.
3. **Horizontal spacing**: select 640.
4. **Vertical spacing**: select 640.
5. You can set the output **Grid** layer to *Create temporary layer* since it is just an intermediate data.

You can then click on *Run*.

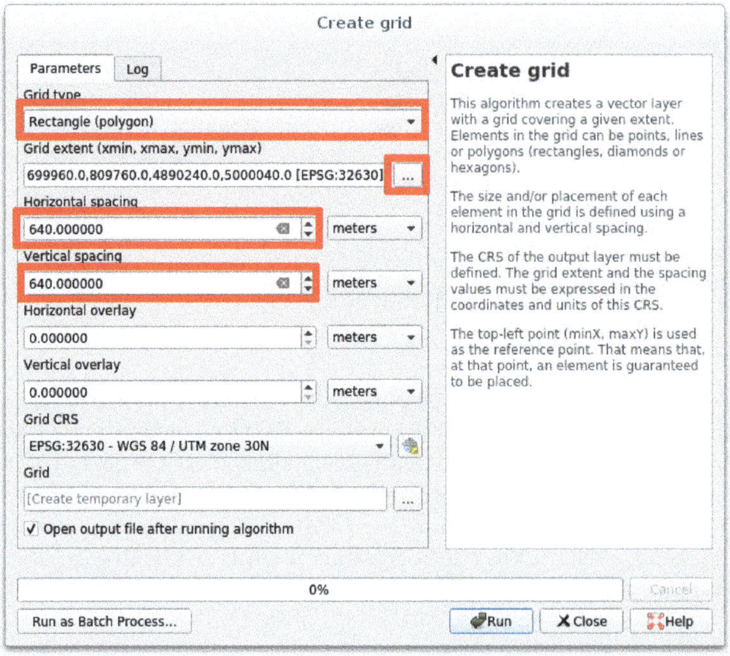

FIGURE 16.10
Vector grid parameters.

16.2.1.4 Grid filtering

We will now keep only the patches of valid pixels. Purposely, we will use the mask created in section 16.2.1.2 to eliminate the undesirable polygons of our grid. Again, vector tools of QGIS are used: click on the menu **Vector** > **Research tools** > **Select by location** (figure 16.11).

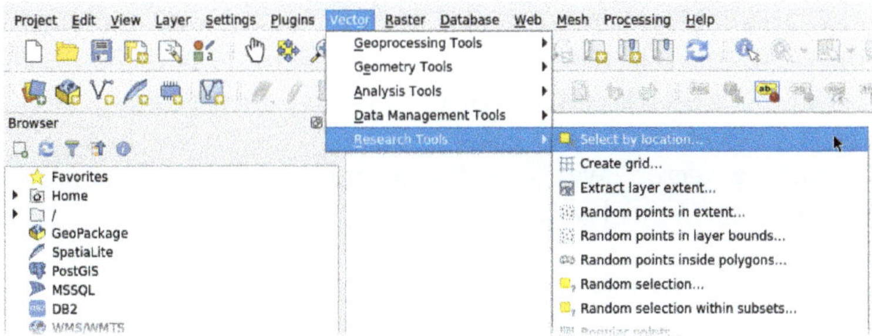

FIGURE 16.11
Select features by location.

In the **Select by location** panel, set the following fields (figure 16.12):

1. **Select features from**: select *Grid*, corresponding to the vector layer that we have generated in the previous step.
2. **By comparing to the features from**: select the *Merged* layer.
3. **Where the features (geometric predicate)**: set *disjoint* since we want to keep the patches that are outside the undesirable areas (clouds and *no-data* values).
4. **Modify current selection by**: select *creating new selection*.

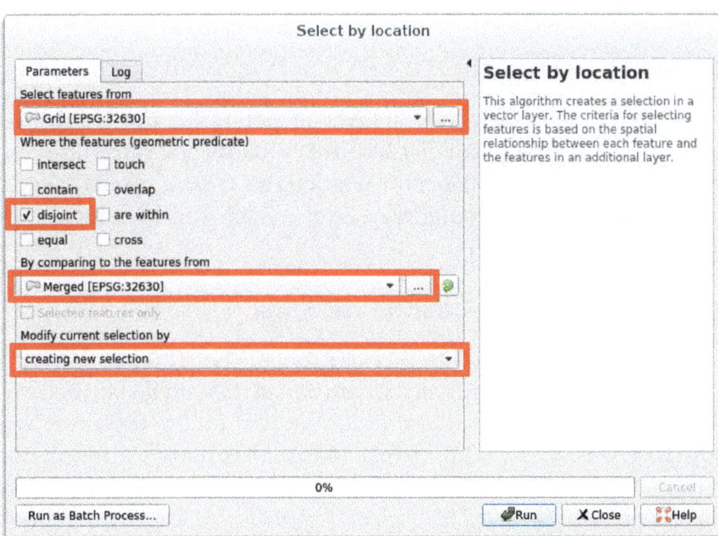

FIGURE 16.12
Select by location.

You can then click on **Run** to generate the selection on the *Grid* vector layer. Then, save the selected features as a new vector layer: right-click on the *Merged* layer and select **Export** > **Save selected features as...** (Fig. 12.7). Save it as *patches_in_valid_areas.gpkg.*

The resulting vector layer should appear as the grid where patches are entirely outside the *holes* layer, as illustrated in figure 16.13.

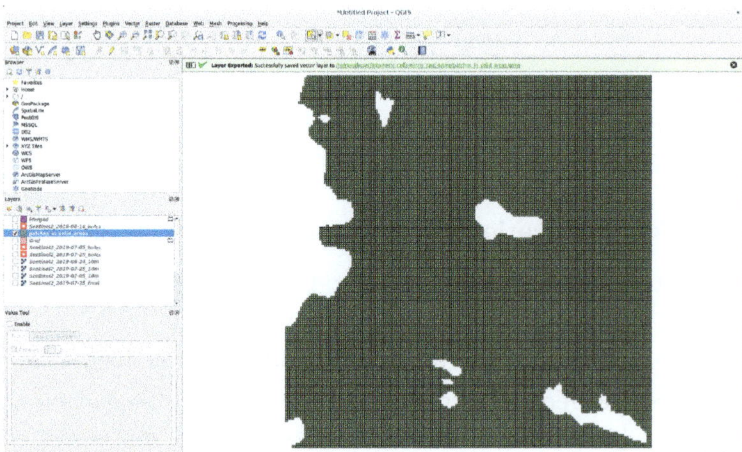

FIGURE 16.13
The resulting layer. From the original grid, only patches that lie completely in valid areas are kept.

16.2.1.5 Patch centroids

Here we simply create a new vector layer of points for the patch centroids. From the menu, click on **Vector** > **Geometry tools** > **Centroids** (as shown in figure 12.5), then extract the centroids for the *Grid* layer. You can let the generated vector layer be a temporary layer, which will be named automatically *Centroids.*

16.2.1.6 Training and validation datasets

Now we need to split the patches in two groups: the *training dataset* will be used to train the network, and the *validation dataset* will be used to assess the performance of the trained network. We will follow the same approach as described in section 12.1.3.3. Select 70% of the patches for the training dataset using from the QGIS menu: select **Vector** > **Research tools** > **Random Selection** (Fig. 12.6). Select *Centroids* as the **input layer**, and select **method** as **percentage of selected features.** Select 70% of the centroids of the layer, and click on **Run** to compute the selection. Then, save the selected features as a new vector layer:

1. Right-click on the *Centroids* layer and select **Export** > **Save selected features as...** (Fig. 12.7). Save it as *patches_centroids_train.gpkg* in your *marmande* directory.
2. From the QGIS menu, click on **Edit**, then **Select**, then **Invert feature selection.**
3. Right-click on the *Centroids* layer and select **Export**, then **Save selected features as...** and save it as *patches_centroids_valid.gpkg* in your *marmande* directory.

You should have now two distinct groups of patch centroids, one for the training dataset and the other for the validation dataset.

Close the .gpkg vector layers in QGIS before processing them. The .gpkg file format for vector layers has an access lock, and .gpkg files cannot be accessed by other software when they are already opened in QGIS. It is thus advised to close the QGIS project, or remove the .gpkg vector layers from the QGIS Layers before running any processing using these files.

16.2.2 Extraction of patches

Now that we know where to extract consistent patches, we proceed to their extraction in the images. We use the **PatchesExtraction** application of OTBTF to perform this step using 4 sources: $S2_{t-1}$, $S2_t$, $S2_{t+1}$, and $S1_t$. In each source, patches of size 64×64 will be extracted. Use the following command line to process the training dataset:

```
export OTB_TF_NSOURCES=4
otbcli_PatchesExtraction \
-vec patches_centroids_train.gpkg \
-source1.il sar_vh_hh_log_norm_EPSG32630.tif \
-source1.patchsizex 64 -source1.patchsizey 64 \
-source1.out patches_train_s1.tif \
-source2.il Sentinel2_2019-07-05_10m.tif \
-source2.patchsizex 64 -source2.patchsizey 64 \
-source2.out patches_train_s2_2019-07-05.tif \
-source3.il Sentinel2_2019-07-25_10m.tif \
-source3.patchsizex 64 -source3.patchsizey 64 \
-source3.out patches_train_s2_2019-07-25.tif \
-source4.il Sentinel2_2019-08-24_10m.tif \
-source4.patchsizex 64 -source4.patchsizey 64 \
-source4.out patches_train_s2_2019-08-24.tif \
-field "id"
```

The **PatchesExtraction** application has created 4 new files that constitute the training dataset:

1. patches_train_s1.tif
2. patches_train_s2_2019-07-05.tif

3. patches_train_s2_2019-07-25.tif
4. patches_train_s2_2019-08-24.tif

Repeat the same operation for the validation dataset, as shown in the following command line.

```
export OTB_TF_NSOURCES=4
otbcli_PatchesExtraction \
-vec patches_centroids_valid.gpkg \
-source1.il sar_vh_hh_log_norm_EPSG32630.tif \
-source1.patchsizex 64 -source1.patchsizey 64 \
-source1.out patches_valid_s1.tif \
-source2.il Sentinel2_2019-07-05_10m.tif \
-source2.patchsizex 64 -source2.patchsizey 64 \
-source2.out patches_valid_s2_2019-07-05.tif \
-source3.il Sentinel2_2019-07-25_10m.tif \
-source3.patchsizex 64 -source3.patchsizey 64 \
-source3.out patches_valid_s2_2019-07-25.tif \
-source4.il Sentinel2_2019-08-24_10m.tif \
-source4.patchsizex 64 -source4.patchsizey 64 \
-source4.out patches_valid_s2_2019-08-24.tif \
-field "id"
```

The **PatchesExtraction** application has created 4 new files that constitute the validation dataset:

1. patches_valid_s1.tif
2. patches_valid_s2_2019-07-05.tif
3. patches_valid_s2_2019-07-25.tif
4. patches_valid_s2_2019-08-24.tif

16.3 More: automate steps with the OTB Python API

In this section, we provide insight on the use of the Python API of the Orfeo ToolBox. This section is, however, optional, and can be skipped since it only shows how to automate the steps previously performed to prepare the patches. The Python API enables you to connect in-memory OTB applications in preserving the so-called streaming mechanism (see 2.1.2). OTB applications can be instantiated as Python objects and their parameters can be set using the same key/value identifiers: once an OTB application object is instantiated, its parameter values can be set using dedicated methods. All parameters that can be set in an OTB application are settable in the Python instance (e.g. integer, float or double precision numbers, strings, images, images lists, vector data, etc.). Hence, it is quite straightforward to implement an entire pipeline that implicates multiple OTB applications chained together: we instantiate

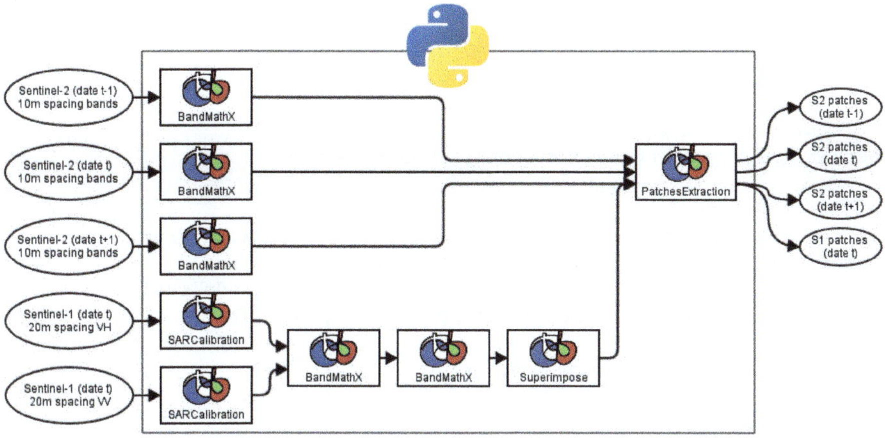

FIGURE 16.14
The pipeline implemented in the Python program. The following applications of the OrfeoToolBox are chained together: **BandMathX**, **SARCalibration**, **Superimpose**, and **PatchesExtraction**.

applications, set their parameters, and connect them together in the desired fashion. After the pipeline is set, one can trigger its execution to generate the outputs.

16.3.1 Build the pipeline

The following Python program automates the pre-processing steps that we saw in this section, apart from the operations performed in QGIS. It consists of chaining together the OTB applications used, from input image calibration and normalization, to the generation of patches. Figure 16.14 illustrates the pipeline implemented in the Python program. The source code follows.

```
import os
import argparse
import otbApplication

parser = argparse.ArgumentParser()

# Input lists of the 10m Sentinel-2 bands (B4, B3, B2 and B8)
parser.add_argument('--img_list_s2_before',required=True,nargs='+',
    default=[])
parser.add_argument('--img_list_s2_t',required=True,nargs='+',
    default=[])
parser.add_argument('--img_list_s2_after', required=True,nargs='+',
    default=[])
```

```
# Input SAR images
parser.add_argument('--img_s1_t_vh', required=True)
parser.add_argument('--img_s1_t_vv', required=True)

# Input vector data
parser.add_argument('--locations', required=True)

# Output patches
parser.add_argument('--out_patches_s2_before', required=True)
parser.add_argument('--out_patches_s2_t', required=True)
parser.add_argument('--out_patches_s2_after', required=True)
parser.add_argument('--out_patches_s1_t', required=True)

params = parser.parse_args()

def normalize_s2(s2_img_list):
  """ This function returns an instance of an OTB application that
      concatenate and normalize the input images
  """
  newApp = otbApplication.Registry.CreateApplication("BandMathX")
  newApp.SetParameterStringList("il", s2_img_list)
  newApp.SetParameterString("exp", "{{{a}*im1b1;{a}*im2b1;{a}*im3b1;
␣␣␣␣␣␣{a}*im4b1}}".format(a="0.00012207"))
  newApp.Execute()
  return newApp

""" Prepare the Sentinel-2 images """

app_s2_norm_1 = normalize_s2(params.img_list_s2_before)
app_s2_norm_2 = normalize_s2(params.img_list_s2_t)
app_s2_norm_3 = normalize_s2(params.img_list_s2_after)

""" Prepare the Sentinel-1 image """

# 1. SAR Calibration
app_s1_vh = otbApplication.Registry.CreateApplication("SARCalibration")
app_s1_vh.SetParameterString("in", params.img_s1_t_vh)
app_s1_vh.Execute()
app_s1_vv = otbApplication.Registry.CreateApplication("SARCalibration")
app_s1_vv.SetParameterString("in", params.img_s1_t_vv)
app_s1_vv.Execute()

# 2. Clip in dB scale
app_s1_vhvv_dB = otbApplication.Registry.CreateApplication("BandMathX")
app_s1_vhvv_dB.AddImageToParameterInputImageList("il",
  app_s1_vh.GetParameterOutputImage("out"))
app_s1_vhvv_dB.AddImageToParameterInputImageList("il",
  app_s1_vv.GetParameterOutputImage("out"))
app_s1_vhvv_dB.SetParameterString("exp",
  "vlog({min(max(im1b1,0.00001),1);min(max(im2b1,0.00001),1)})")
app_s1_vhvv_dB.Execute()

# 3. Normalize between [-1, 0]
app_s1_vhvv_dB_norm = otbApplication.Registry.CreateApplication
```

```
  ("BandMathX")
app_s1_vhvv_dB_norm.AddImageToParameterInputImageList("il",
  app_s1_vhvv_dB.GetParameterOutputImage("out"))
app_s1_vhvv_dB_norm.SetParameterString("exp",
  "{im1b1/im1b1Mini;␣im1b2/im1b2Mini}")
app_s1_vhvv_dB_norm.Execute()

# 4. Superimpose on optical images
app_s1_resampled = otbApplication.Registry.CreateApplication
  ("Superimpose")
app_s1_resampled.SetParameterInputImage("inr",
  app_s2_norm_1.GetParameterOutputImage("out"))
app_s1_resampled.SetParameterInputImage("inm",
  app_s1_vhvv_dB_norm.GetParameterOutputImage("out"))
app_s1_resampled.Execute()

""" Patches extraction """

# Tell the PatchesExtraction application that we want 4 sources
os.environ['OTB_TF_NSOURCES'] = "4"
app_patches = otbApplication.Registry.CreateApplication
  ("PatchesExtraction")
app_patches.AddImageToParameterInputImageList("source1.il",
  app_s2_norm_1.GetParameterOutputImage("out"))
app_patches.SetParameterString("source1.out", params.out_patches_
  s2_before)
app_patches.AddImageToParameterInputImageList("source2.il",
  app_s2_norm_2.GetParameterOutputImage("out"))
app_patches.SetParameterString("source2.out", params.out_patches_
  s2_t)
app_patches.AddImageToParameterInputImageList("source3.il",
  app_s2_norm_3.GetParameterOutputImage("out"))
app_patches.SetParameterString("source3.out", params.out_patches_s2_
  after)
app_patches.AddImageToParameterInputImageList("source4.il",
  app_s1_resampled.GetParameterOutputImage("out"))
app_patches.SetParameterString("source4.out", params.out_patches_s1_t)

# Set all sources patches sizes to 64
for i in [1,2,3,4]:
  for dim in ["x", "y"]:
    app_patches.SetParameterInt("source{}.patchsize{}".format(i, dim),
       64)

# The input vector data must have an "id" field
app_patches.SetParameterString("vec", params.locations)
app_patches.SetParameterString("field", "id")
app_patches.ExecuteAndWriteOutput()
```

16.3.2 Run the pipeline

Below is a command line to run the program to generate the training patches. To generate the validation patches, it suffices to replace the name of the input vector data, and change the names for the output patches.

```
# Run the pipeline to generate the patches for training
python preprocess_gapfill.py \
--locations patches_centroids_train.gpkg \
--img_list_s2_before \
  T30TYQ_20190705T105031_B04_10m.jp2 \
  T30TYQ_20190705T105031_B03_10m.jp2 \
  T30TYQ_20190705T105031_B02_10m.jp2 \
  T30TYQ_20190705T105031_B08_10m.jp2 \
--img_list_s2_t \
  T30TYQ_20190725T105031_B04_10m.jp2 \
  T30TYQ_20190725T105031_B03_10m.jp2 \
  T30TYQ_20190725T105031_B02_10m.jp2 \
  T30TYQ_20190725T105031_B08_10m.jp2 \
--img_list_s2_after \
  T30TYQ_20190824T105031_B04_10m.jp2 \
  T30TYQ_20190824T105031_B03_10m.jp2 \
  T30TYQ_20190824T105031_B02_10m.jp2 \
  T30TYQ_20190824T105031_B08_10m.jp2 \
--img_s1_t_vh \
  s1a-iw-grd-vh-20190725t175553-20190725t175618-028277-0331c8-002.tiff \
--img_s1_t_vv \
  s1a-iw-grd-vv-20190725t175553-20190725t175618-028277-0331c8-001.tiff \
--out_patches_s2_before patches_train_s2_2019-07-05.tif \
--out_patches_s2_t patches_train_s2_2019-07-25.tif \
--out_patches_s2_after patches_train_s2_2019-08-24.tif \
--out_patches_s1_t patches_train_s1.tif
```

17

Model training

This section addresses model training. We will follow the following steps to get our model trained and ready, in order to perform gapfilling:

1. Download the code of the implementation from the author's repository.
2. Use the code to train the model. Model weights will be saved in *checkpoint* files.
3. Export a trained ***SavedModel*** from an existing *checkpoint.*
4. Use the ***SavedModel*** with the OTBTF **TensorflowModelServe** application to perform the optical image restoration.

17.1 Training from Python

Many practitioners like to have full control of their training process, and will prefer the TensorFlow Python API rather than using the **TensorflowModelTrain** application. Indeed, **TensorflowModelTrain** hides the complexity of the training and provides the user an easy interface to train models on their geospatial data, but with a drastically simplified training approach. However, this user-friendliness comes with the cost of over-simplifying the process, which can be frustrating for deep learning experts and people that would prefer a more customized approach to train their models. In this section, we show that experts can still train their models from their own Python code, and apply the trained model using OTBTF. The *deep-gapfill* code illustrates this in providing the build and training of a deep learning architecture in pure Python. The resulting model can be exported as a TensorFlow ***SavedModel*** and used with the **TensorflowModelServe** application to generate the gap-filled image.

17.2 Get the code

First, download the Python code of the implementation described in section 14.3. You can clone the source code from the author's github account (*deep-gapfill* repository[1]) or directly download the archive of the code.[2] The *deep-gapfill* code enables you to directly train the model.

17.3 Use the code

17.3.1 Description

From your command line interface, navigate to the folder where you have downloaded *deep-gapfill.* You should have a *model.py* file that you will launch with Python:

```
python model.py
```

The application should display the help function. From the displayed description, you should be able to identify the important inputs and parameters that will be used to train the model:

1. **s1_t**, **s2_t**, **s2_t_before** and **s2_t_after** are the paths to the input patch images for training, respectively, for $S1_t$, $S2_t$, $S2_{t-1}$ and $S2_{t+1}$.
2. **valid_s1_t**, **valid_s2_t**, **valid_s2_t_before** and **valid_s2_t_after** are the paths to the input patch images for validation, respectively, for $S1_t$, $S2_t$, $S2_{t-1}$ and $S2_{t+1}$.
3. **save_ckpt** is the parameter for the output prefix of *checkpoint* files, that are used by TensorFlow to save the model weights on disk.

You can notice that the inputs for $S1_t$ are not mandatory: when **s1_t** and **valid_s1_t**, aren't provided, the model will be built without the SAR encoder branch. This ablation feature enables you to compare how the model behaves with or without the SAR information.

17.3.2 Parameters

The table 17.1 summarizes the parameter values that will be used to run the training:

[1] `https://github.com/remicres/Deep-Gapfill`
[2] `https://github.com/remicres/Deep-Gapfill/archive/master.zip`

TABLE 17.1
model.py parameters

Parameter	Value used
s1_t	patches_train_s1.tif
s2_t	patches_train_s2_2019-07-05.tif
s2_t_before	patches_train_s2_2019-07-25.tif
s2_t_after	patches_train_s2_2019-08-24.tif
valid_s1_t	patches_valid_s1.tif
valid_s2_t	patches_valid_s2_2019-07-05.tif
valid_s2_t_before	patches_valid_s2_2019-07-25.tif
valid_s2_t_after	patches_valid_s2_2019-08-24.tif

Run the Python code from the directory containing the results, using the following command.

```
# clone the repository (if you don't have it already)
git clone https://github.com/remicres/Deep-Gapfill.git

# run the model training
python Deep-Gapfill/model.py \
--s1_t patches_train_s1.tif \
--s2_t_before patches_train_s2_2019-07-05.tif \
--s2_t patches_train_s2_2019-07-25.tif \
--s2_t_after patches_train_s2_2019-08-24.tif \
--valid_s1_t patches_valid_s1.tif \
--valid_s2_t_before patches_valid_s2_2019-07-05.tif \
--valid_s2_t patches_valid_s2_2019-07-25.tif \
--valid_s2_t_after patches_valid_s2_2019-08-24.tif \
--save_ckpt mymodel_checkpoint
```

The program will execute and achieve the following actions:

1. Load the input patch images.
2. For each epoch:
   (a) Perform the gradient descent over the dataset.
   (b) Assess the model error (in terms of L1 and L2 losses).
   (c) Save the model weights into a *checkpoint* file.

17.4 Export the model

Once all epochs are completed, we can export a ***SavedModel*** from the last *checkpoint* created during the training. Note that a better approach could be to export the best model trained among the last epochs, or as soon as the model over-fit the training data (there are multiple strategies, and we do not intend to cover this topic in this book). To export the ***SavedModel***, use the Python code located in the *Python* folder of OTBTF named *ckpt2savedmodel.py*.

```
# Convert the checkpoint into a SavedModel
python /work/otb/otb/Modules/Remote/otbtf/python/ckpt2savedmodel.py \
--ckpt mymodel_checkpoint-199 \
--inputs "s1:0" "s2_before:0" "s2_after:0" \
--outputs "gen_fcn:0" \
--model savedmodel_gapfill
```

After the command run, a new folder that contains the ***SavedModel*** should have been created (*savedmodel_gapfill*).

18

Inference

We will apply the previously exported ***SavedModel*** on the Sentinel-1 and Sentinel-2 images.

18.1 Inputs and outputs

At inference time, only the inputs for $S1_t$, $S2_{t-1}$ and $S2_{t+1}$ are fed to the model. The $\widehat{S2_t}$ input is not used anymore: we only care about retrieving the $S2_t$ image at the output of the model.

Another important point is the output used to retrieve $S2_t$. As explained in section 12.2.2.1, computing the *exact* output of a fully convolutional model is one strategy to avoid blocking artifacts. For this purpose, we compute the *exact* output of the model in extracting the center of the final convolution output. We then remove 256 pixels from each side of the output tensor named *prediction* in the spatial dimensions. The corresponding tensor, named *prediction_fcn*, is used as output for $\widehat{S2_t}$. Removing 256 pixels from each side of the spatial dimensions of the output tensor is equivalent to using a model that inputs virtual patches of 576×576 and produces a field of 64×64 pixels. Note that very deep models can lead to a significant theoretical margin to remove. When this margin is too large, one is advised to analyze the effect of the margin on the output, to find a practical setting. The practical margin should be large enough to avoid blocking artifacts, and small enough to reasonably impact the shrinking of the model output. In the OTBTF formalism, our network has hence a receptive field of 576×576 pixels, and an output *expression field* of 64×64 pixels. Figure 18.1 presents the model as used at inference time.

18.2 Generating the image

Since the output is *prediction_fcn*, the image can be efficiently processed by blocks without tiling artifacts. We apply the model on the inputs

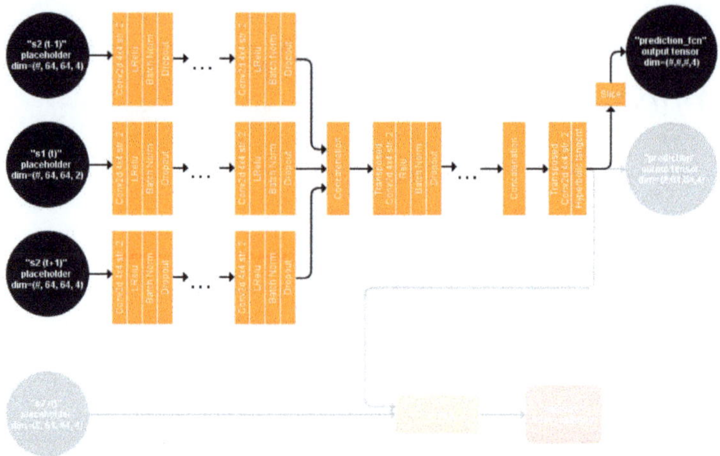

FIGURE 18.1
Only $S1_t$, $S2_{t-1}$ and $S2_{t+1}$ inputs are used. Nodes related to training (cost function, optimizer) are not used anymore at inference time. The $S2_t$ image is retrieved at the output of the model from the *prediction_fcn* tensor. *prediction_fcn* is the *exact* output of the model, obtained in extracting the center of the *prediction* tensor in the spatial dimensions.

$S2_{t-1}$, $S2_{t+1}$ and $S1_t$ and generate the output image $\widehat{S2_t}$ that we name *Sentinel2_2019-07-25_estimated.tif* Run the following command line to generate the synthetic $\widehat{S2_t}$ image:

```
export OTB_TF_NSOURCES=3
otbcli_TensorflowModelServe \
-source1.il sar_vh_hh_log_norm_EPSG32630.tif \
-source1.rfieldx 576 -source1.rfieldy 576 \
-source1.placeholder "s1" \
-source2.il Sentinel2_2019-07-05_10m.tif \
-source2.rfieldx 576 -source2.rfieldy 576 \
-source2.placeholder "s2_before" \
-source3.il Sentinel2_2019-08-24_10m.tif \
-source3.rfieldx 576 -source3.rfieldy 576 \
-source3.placeholder "s2_after" \
-model.dir savedmodel_gapfill \
-model.fullyconv on \
-output.names "gen_fcn" -output.efieldx 64 -output.efieldy 64 \
-out Sentinel2_2019-07-25_estimated.tif
```

Once the image is generated, import it in QGIS and compare it visually with the $S2_t$ image (figure 18.2).

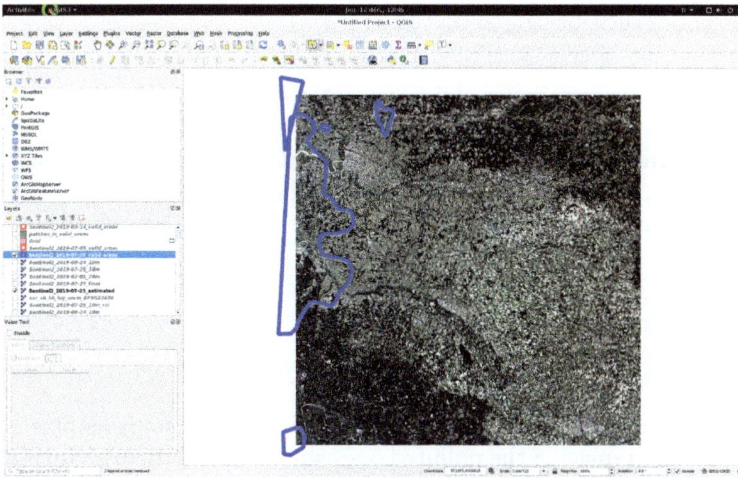

FIGURE 18.2
The resulting synthetic image. Note that the pixels are cloud-free in the areas which were polluted in the original $S2_t$ image (blue polygons).

> **Speeding up the process**
> In order to accelerate the process, we recommend setting the *optim.gridspacingx/y* to larger values (default is 64 in both axes, because it is the smallest multiple of the *expression field* width/height value). Depending on the available hardware, tiles of 512, 1024 or 2048 can be generated. The larger the *optim.gridspacingx/y* is, the faster the output image will be generated. However, the hardware must have enough memory to efficiently perform the computations.

18.3 Postprocessing

We can keep the original values of $S2_t$ where the pixels were not polluted by clouds. To perform this, the cloudy part of the $S2_t$ image is replaced with the $\widehat{S2}_t$ image in the areas designated with the *Sentinel2_2019-07-25_holes.gpkg* vector layer. First we create a binary mask image representing the area that we want to replace in the $S2_t$ image. Using the **Rasterization** OTB application, we create a binary image from the *Sentinel2_2019-07-25_holes.gpkg* vector layer. This binary image will have the same size as the synthetic Sentinel-2 image.

```
otbcli_Rasterization \
-in Sentinel2_2019-07-25_holes.gpkg \
```

```
-im Sentinel2_2019-07-25_estimated.tif \
-out valid_area_mask.tif uint8
```

After the rasterization is done, we interpolate the $S2_t$ image over the $\widehat{S2_t}$ image. Because of the receptive field, the generated $\widehat{S2_t}$ image is slightly smaller than the $S2_t$ image. In order to conduct a pixel-wise operation on pixels of both $S2_t$, $\widehat{S2_t}$ and the binary mask, we must superimpose the $S2_t$ image over the $\widehat{S2_t}$ image. To perform this operation, we use the **Superimpose** OTB application:

```
otbcli_Superimpose \
-inr Sentinel2_2019-07-25_estimated.tif \
-inm Sentinel2_2019-07-25_10m.tif \
-out Sentinel2_2019-07-25_10m_superimposed.tif
```

We can then compose the final Sentinel-2 image, keeping the pixels in the valid areas, and replacing the polluted ones with $\widehat{S2_t}$. The **BandMathX** OTB application allows us to create the final image. A mathematical expression is used to perform the computation of the output pixel. **BandMathX** is able to identify the input images and the input image bands in a mathematical expression. The **BandMathX** expression is the following:

im1b1==255 ? im2 : im3

The "?" and ":" are ternary operators. The expression is interpreted as: If the binary mask (i.e. *im1b1*: like "band #1 of image #1") is 255, then the pixel is polluted and the output is the $S2_t$ image (i.e. *im2*: like "image #2"). Otherwise, the output is the $\widehat{S2_t}$ image (i.e. *im3*: like "image #3"). Run the following command line to perform the final compositing step of the gapfilling:

```
otbcli_BandMathX \
-il valid_area_mask.tif \
    Sentinel2_2019-07-25_estimated.tif \
    Sentinel2_2019-07-25_10m_superimposed.tif \
-exp "im1b1==255␣?␣im2␣:␣im3" \
-out Sentinel2_2019-07-25_final.tif
```

Note that at this step, we can also instead denormalize the image in inverting the equation 16.1.1, i.e. multiplying $im2$ and $im3$ by 8192:

```
otbcli_BandMathX \
-il valid_area_mask.tif \
    Sentinel2_2019-07-25_estimated.tif \
    Sentinel2_2019-07-25_10m_superimposed.tif \
-exp "im1b1==255␣?␣8192*im2␣:␣8192*im3" \
-out Sentinel2_2019-07-25_final_16b.tif int16
```

The final image is ready. Figure 18.3 shows the resulting image imported in QGIS.

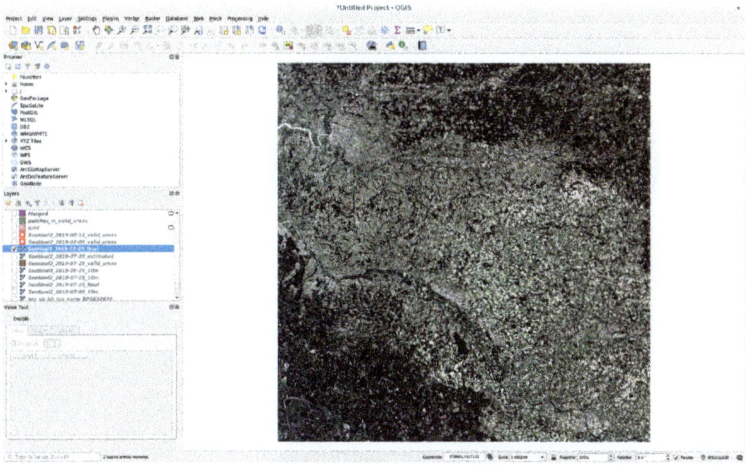

FIGURE 18.3
The final image resulting from the gapfilling process.

19

Discussion

This section introduced how to produce one synthetic remote sensing image (i) from multiple sources, and (ii) in making use directly of TensorFlow to train the model from Python code. We have applied a method which performs the gapfilling of a cloudy optical image, from multiple optical images and a synthetic aperture radar image.

The dataset used during this exercise consisted of three optical Sentinel-2 images, including one scene polluted by clouds, and one SAR Sentinel-1 image. After we carefully rejected the cloudy parts of the images in the patch selection, we proceeded to the patch extraction to construct two datasets: one for the training, the other for validation.

Once the patch extraction was done, we trained our model. We ran a Python code that directly employs TensorFlow to perform the training, using the extracted patches of the four Sentinel images. We trained a deep convolutional neural network that works as an autoencoder, to estimate the cloud-free optical Sentinel-2 image from the two other optical images (acquired respectively before and after the cloudy one), and the SAR image.

After our model was trained, we exported it in the ***SavedModel*** format. We applied the trained model to generate an entire synthetic optical Sentinel-2 image corresponding to the polluted scene. In order to combine the genuine cloudy Sentinel-2 image with the synthetic one, we used the masks to compose the final image from both the synthetic (i.e. generated with the deep learning model) and the original one (i.e. the parts of the image that were not polluted by clouds).

It should be noted that the images acquired respectively before and after the cloudy images are not completely free of clouds: one could extend this exercise to the retrieval of the missing contents in each of these images, repeating the steps described in this section with each image as a target estimated image, one by one.

Another interesting extension of this exercise could be to modify the model to use the 20-m spacing channels of the input optical images (spectral bands 5, 6, 7, 8a, 11 and 12 for Sentinel-2). The model should be able to estimate also the 20-m spacing channels of the optical images, using a second output. For instance: in the encoder, we could add new inputs for the 20-m optical images just after the first strided convolution: this would enable the fusion of the first feature map (the one that is computed from the 10-m spacing input images using stride 2 convolution, hence having 20-m physical spacing) with the input image at 20 m.

Another useful modification of the model could be to integrate the image normalization, avoiding some pre-processing, and saving storage.

The introduced application, optical time series gapfilling using a deep convolutional autoencoder, has illustrated how a deep neural network can easily combine multiple sources having different modalities, physical and spectral resolutions. While we have illustrated the versatility of deep learning through this particular application, we have no doubt that a huge number of problems in remote sensing can be addressed with carefully tailored deep learning architectures.

Bibliography

[1] I. Goodfellow, Y. Bengio, and A. Courville, *Deep Learning*. MIT Press, 2016.

[2] S. Chen and H. Wang, "SAR target recognition based on deep learning," in *Data Science and Advanced Analytics (DSAA), 2014 International Conference on*, pp. 541–547, IEEE, 2014.

[3] D. H. T. Minh, D. Ienco, R. Gaetano, N. Lalande, E. Ndikumana, F. Osman, and P. Maurel, "Deep recurrent neural networks for winter vegetation quality mapping via multitemporal SAR sentinel-1," *IEEE Geoscience and Remote Sensing Letters*, vol. 15, no. 3, pp. 464–468, 2018.

[4] L. Wang, K. A. Scott, L. Xu, and D. A. Clausi, "Sea ice concentration estimation during melt from dual-pol SAR scenes using deep convolutional neural networks: A case study," *IEEE Transactions on Geoscience and Remote Sensing*, vol. 54, no. 8, pp. 4524–4533, 2016.

[5] Y. Chen, Z. Lin, X. Zhao, G. Wang, and Y. Gu, "Deep learning-based classification of hyperspectral data," *IEEE Journal of Selected Topics in Applied Earth Observations and Remote Sensing*, vol. 7, pp. 2094–2107, June 2014.

[6] W. Li, G. Wu, and Q. Du, "Transferred deep learning for anomaly detection in hyperspectral imagery," *IEEE Geoscience and Remote Sensing Letters*, vol. 14, pp. 597–601, May 2017.

[7] M. Volpi and D. Tuia, "Dense semantic labeling of subdecimeter resolution images with convolutional neural networks," *IEEE Transactions on Geoscience and Remote Sensing*, vol. 55, no. 2, pp. 881–893, 2017.

[8] E. Maggiori, Y. Tarabalka, G. Charpiat, and P. Alliez, "Convolutional neural networks for large-scale remote-sensing image classification," *IEEE Transactions on Geoscience and Remote Sensing*, vol. 55, no. 2, pp. 645–657, 2017.

[9] G. Cheng and J. Han, "A survey on object detection in optical remote sensing images," *ISPRS Journal of Photogrammetry and Remote Sensing*, vol. 117, pp. 11–28, 2016.

[10] P. Napoletano, "Visual descriptors for content-based retrieval of remote-sensing images," *International Journal of Remote Sensing*, vol. 39, no. 5, pp. 1343–1376, 2018.

[11] D. Ienco, R. Gaetano, C. Dupaquier, and P. Maurel, "Land cover classification via multitemporal spatial data by deep recurrent neural networks," *IEEE Geoscience and Remote Sensing Letters*, vol. 14, no. 10, pp. 1685–1689, 2017.

[12] L. Gomez-Chova, D. Tuia, G. Moser, and G. Camps-Valls, "Multimodal classification of remote sensing images: A review and future directions," *Proceedings of the IEEE*, vol. 103, no. 9, pp. 1560–1584, 2015.

[13] P. Benedetti, D. Ienco, R. Gaetano, K. Osé, R. Pensa, and S. Dupuy, "M3fusion: A deep learning architecture for multi-{Scale/Modal/Temporal} satellite data fusion," *arXiv preprint arXiv:1803.01945*, 2018.

[14] G. Masi, D. Cozzolino, L. Verdoliva, and G. Scarpa, "Pansharpening by convolutional neural networks," *Remote Sensing*, vol. 8, no. 7, p. 594, 2016.

[15] L. Zhang, L. Zhang, and B. Du, "Deep learning for remote sensing data: A technical tutorial on the state of the art," *IEEE Geoscience and Remote Sensing Magazine*, vol. 4, no. 2, pp. 22–40, 2016.

[16] M. Grizonnet, J. Michel, V. Poughon, J. Inglada, M. Savinaud, and R. Cresson, "Orfeo ToolBox Open source processing of remote sensing images," *Open Geospatial Data, Software and Standards*, vol. 2, no. 1, p. 15, 2017.

[17] M. Abadi, P. Barham, J. Chen, Z. Chen, A. Davis, J. Dean, M. Devin, S. Ghemawat, G. Irving, M. Isard, *et al.*, "Tensorflow: A system for large-scale machine learning," in *12th {USENIX} Symposium on Operating Systems Design and Implementation ({OSDI} 16)*, pp. 265–283, 2016.

[18] R. Cresson, "A framework for remote sensing images processing using deep learning techniques," *IEEE Geoscience and Remote Sensing Letters*, vol. 16, pp. 25–29, 2019.

[19] D. P. Kingma and J. Ba, "Adam: A method for stochastic optimization," *arXiv preprint arXiv:1412.6980*, 2014.

[20] K. Nogueira, O. A. Penatti, and J. A. dos Santos, "Towards better exploiting convolutional neural networks for remote sensing scene classification," *Pattern Recognition*, vol. 61, pp. 539–556, 2017.

[21] D. Marmanis, M. Datcu, T. Esch, and U. Stilla, "Deep learning earth observation classification using ImageNet pretrained networks," *IEEE Geoscience and Remote Sensing Letters*, vol. 13, no. 1, pp. 105–109, 2016.

[22] F. Hu, G.-S. Xia, J. Hu, and L. Zhang, "Transferring deep convolutional neural networks for the scene classification of high-resolution remote sensing imagery," *Remote Sensing*, vol. 7, no. 11, pp. 14680–14707, 2015.

[23] R. Gaetano, D. Ienco, K. Ose, and R. Cresson, "MRFusion: A deep learning architecture to fuse PAN and MS imagery for land cover mapping," *arXiv preprint arXiv:1806.11452*, 2018.

[24] K. Simonyan and A. Zisserman, "Very deep convolutional networks for large-scale image recognition," *arXiv preprint arXiv:1409.1556*, 2014.

[25] O. Ronneberger, P. Fischer, and T. Brox, "U-Net: Convolutional networks for biomedical image segmentation," in *International Conference on Medical Image Computing and Computer-Assisted Intervention*, pp. 234–241, Springer, 2015.

[26] L.-C. Chen, G. Papandreou, I. Kokkinos, K. Murphy, and A. L. Yuille, "DeepLab: Semantic image segmentation with deep convolutional nets, atrous convolution, and fully connected CRFs," *IEEE Transactions on Pattern Analysis and Machine Intelligence*, vol. 40, no. 4, pp. 834–848, 2017.

[27] H. Zhao, J. Shi, X. Qi, X. Wang, and J. Jia, "Pyramid scene parsing network," in *Proceedings of the IEEE Conference on Computer Vision and Pattern Recognition*, pp. 2881–2890, 2017.

[28] J. Inglada, M. Arias, B. Tardy, O. Hagolle, S. Valero, D. Morin, G. Dedieu, G. Sepulcre, S. Bontemps, P. Defourny, and B. Koetz, "Assessment of an operational system for crop type map production using high temporal and spatial resolution satellite optical imagery," *Remote Sensing*, vol. 7, no. 9, pp. 12356–12379, 2015.

[29] M. Xu, X. Jia, M. Pickering, and A. J. Plaza, "Cloud removal based on sparse representation via multitemporal dictionary learning," *IEEE TGRS*, vol. 54, no. 5, pp. 2998–3006, 2016.

[30] X. Li, H. Shen, L. Zhang, H. Zhang, Q. Yuan, and G. Yang, "Recovering quantitative remote sensing products contaminated by thick clouds and shadows using multitemporal dictionary learning," *IEEE TGRS*, vol. 52, no. 11, pp. 7086–7098, 2014.

[31] K. Enomoto, K. Sakurada, W. Wang, H. Fukui, M. Matsuoka, R. Nakamura, and N. Kawaguchi, "Filmy cloud removal on satellite imagery with multispectral conditional generative adversarial nets," in *CVPRw*, pp. 1533–1541, 2017.

[32] C. Grohnfeldi, M. Schmitt, and X. Zhu, "A conditional generative adversarial network to fuse sar and multispectral optical data for cloud removal from sentinel-2 images," in *IGARSS*, pp. 1726–1729, IEEE, 2018.

[33] G. Scarpa, M. Gargiulo, A. Mazza, and R. Gaetano, "A CNN-based fusion method for feature extraction from sentinel data," *Remote Sensing*, vol. 10, no. 2, p. 236, 2018.

[34] R. Cresson, D. Ienco, R. Gaetano, K. Ose, and D. H. T. Minh, "Optical image gap filling using deep convolutional autoencoder from optical and radar images," in *IGARSS 2019-2019 IEEE International Geoscience and Remote Sensing Symposium*, pp. 218–221, IEEE, 2019.

[35] S. Ioffe and C. Szegedy, "Batch normalization: Accelerating deep network training by reducing internal covariate shift," *arXiv:1502.03167*, 2015.

Index

For Product Safety Concerns and Information please contact our EU representative GPSR@taylorandfrancis.com
Taylor & Francis Verlag GmbH, Kaufingerstraße 24, 80331 München, Germany

www.ingramcontent.com/pod-product-compliance
Ingram Content Group UK Ltd.
Pitfield, Milton Keynes, MK11 3LW, UK
UKHW022211190926
475702UK00003B/306
* 9 7 8 0 3 6 7 8 5 8 4 8 3 *